Materials Technology 4

Materials Technology 4

W. Bolton

Senior Adviser, Technician Education Council

First published 1981
Reprinted 1985

British Library Cataloguing in Publication Data

Bolton, W.
Materials technology 4.
1. Materials science
I. Title
620.1′1092′062 TA403

ISBN 0-408-00584-X

Typeset by Tunbridge Wells Typesetting Services
Printed and bound by
The Whitefriars Press Ltd, London and Tonbridge

Acknowledgements

Thanks are due to the following sources for permitting the reproduction of items from their publications and for their assistance in the preparation of the book.

Alcan Extrusions Ltd.
Aluminium Extruders Association
Bayer UK Ltd
BCIRA
British Industrial Plastics Ltd.
British Steel Corporation, Tubes Division
BSC Stainless
Darlington & Simpson Rolling Mills Ltd.
Darwins Alloy Castings Ltd.
Deloro Stellite
Delta Extruded Metals Co. Ltd.
Du Pont (UK) Ltd.
Dynacast International Ltd
Engineering
Fothergill and Harvey Ltd
GKN Steelstock Ltd.
Hodder and Stoughton Ltd.
Holcroft Castings and Forgings Ltd.
Inco Europe Ltd.
Arthur Lee & Sons Ltd.
Lee Bright Bars Ltd.
Longman Group Ltd.
Macmillan & Co. Ltd.
McKechnie Metals Ltd.
New Scientist
Osborn Steel Extrusions
Sterling Metals Ltd.
The Design Council
Tufnol Ltd.
Wokingham Plastics Ltd.

Preface

This book has been written with the following ,

1. To provide a basic knowledge of the way when in use (Chapters 1 to 5).
2. To provide an overview of the forming an used in engineering (Chapters 6 to 8).
3. To consider the problems of material and for components (Chapter 9 and 10).

The book covers the unit Materials Technolog U78/477) of the Technician Education Council. to have a subtitle it would be 'an introduction to materials and processes'. The unit forms part of earlier units being concerned with basic mechanic structures of materials and the testing procedures in engineering. Chapters 1 and 5 of this book incl of revision of this earlier material.

Contents

1 Basic mechanical properties

Objectives: At the end of this chapter you should be able to:
Explain the terms modulus of elasticity, tensile strength, compressive strength, elastic limit, percentage elongation, yield stress, proof stress.
Describe the form of typical stress/strain graphs for brittle and ductile metals, plastics and rubbers.
Explain how hardness is measured.
Explain the two principal forms of impact testing and the significance of the results of such tests.
Compare the mechanical properties of metals, polymers, ceramics and composites.
Explain the elastic and plastic behaviour of metals qualitatively by a simple dislocation model and plastics by a simple consideration of molecule chains.
Interpret data given for materials properties.

STRESS/STRAIN GRAPHS

The behaviour of materials subject to tensile and compressive forces can be described in terms of their stress/strain behaviour, stress being the applied force per unit area and strain the extension, or contraction, per unit original length of material. The area referred to in the stress definition is the area at right-angles to the line of action of the force. In stress/strain graphs it is usually the area before any forces are applied, no account thus being taken of the reduction in area that occurs when the material is stretched or of the increase in area if it is compressed.

Figure 1.1 shows stress/strain graphs for typical samples of mild steel and cast iron. For both materials there is a part of the graph where the strain is directly proportional to the stress and for this region a *modulus of elasticity* is defined as

$$\text{modulus of elasticity } E = \frac{\text{stress}}{\text{strain}}$$

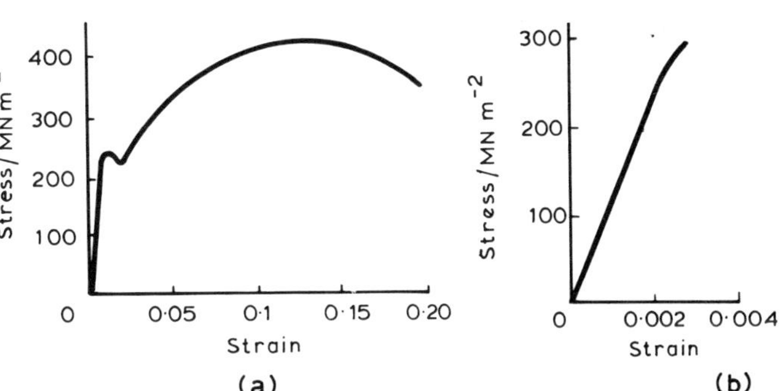

Figure 1.1 Stress-strain graphs for (a) mild steel (b) cast iron

This modulus is often referred to as Young's modulus and is a measure of the ease of stretching or compressing a material i.e. the stiffness. The higher the value of the modulus the more stress is needed to produce a given extension. Some typical values for metals are:

Material	*Modulus of elasticity*/GN m^{-2} *or* kN mm^{-2}
Mild steel	220
Cast iron	150
Brass	120
Aluminium alloy	70

Thus, less stiff aluminium alloy is easier to extend than mild steel. For most materials the modulus of elasticity is the same in tension as in compression.

The maximum tensile stress that a material can withstand is known as the *tensile strength.* In the case of the mild steel in *Figure 1.1a,* the tensile strength is about 400 MN m^{-2}, for the cast iron in *Figure 1.1b* about 300 MN m^{-2}. Some typical values for metals are:

Material	*Tensile strength*/MN m^{-2} *or* N mm^{-2}
Mild steel	400
Cast iron	140–300
Brass	120–400
Aluminium alloy	140–600

The maximum stress which materials can withstand in compression is known as the *compressive strength,* which is not usually the same as the tensile strength.

When a material has a stress applied to it a strain results. When the stress is removed the strain can vanish and is said to have been *elastic* strain. Up to some limiting stress, most materials will exhibit elastic strain, the limiting stress being called the *elastic limit.* Stresses above this limit give rise to a permanent set, i.e. the material does not return to its original length when the stress is removed. Above this limiting stress the material is said to exhibit a combination of elastic and plastic behaviour, a combination because, generally, not all the strain is retained as a permanent set.

For some materials the difference between the elastic limit stress and the stress at which failure occurs is very small, as with the cast iron in *Figure 1.1b.* The length of the sample after breaking is not much different from the initial length of the sample, very little permanent set having occurred. Such materials are said to be *brittle,* in contrast to a material which suffers a considerable amount of plastic strain before breaking, which is called a *ductile* material, e.g. the mild steel in *Figure 1.1a.* A measure of ductility is the *percentage elongation* of the sample after breaking; a very ductile material can have a percentage elongation at breaking of 40 per cent, a brittle material perhaps just a few per cent.

With some materials, e.g. the mild steel in *Figure 1.1a*, there is a noticeable dip in the stress/strain graph at some specific stress, where the strain increases without any increase in load. The material is said to have yielded and the stress at which this occurs is called the *yield stress.* Some materials, such as aluminium alloys, do not show a noticeable yield stress point, and it is usual here to specify a 0.2 per

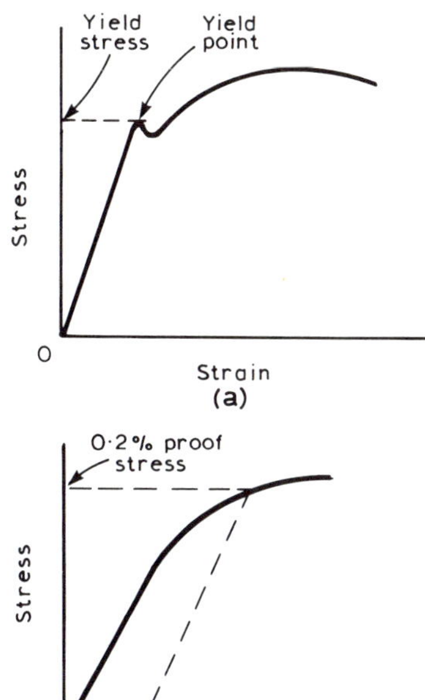

Figure 1.2 Yield and proof stress

cent *proof stress* where 0.2 is the percentage off-set (*Figure 1.2*). Some typical values of yield and proof stress for metals are:

Material	*Yield stress* /MN m^{-2} *or* N mm^{-2}	*0.2% Proof stress* /MN m^{-2} *or* N mm^{-2}
Mild steel	230	–
Stainless steel	620	–
Brass, drawn	–	200
Manganese bronze	–	250
Copper	–	240

In general all the above discussions concerning the stress/strain graphs for metals can be applied to those for plastics. *Figure 1.3* shows a stress/strain graph for ABS polymer. The stress/strain

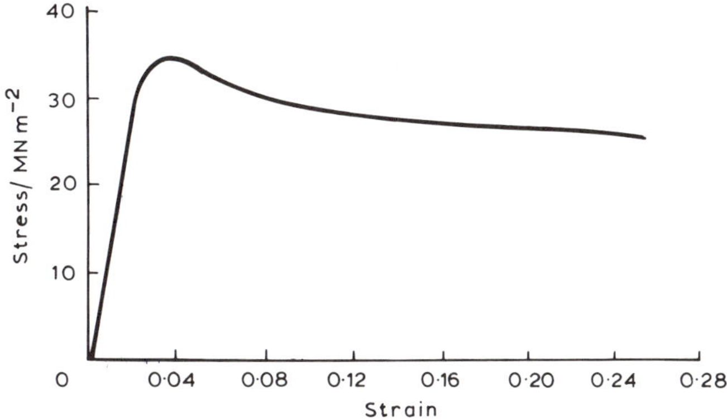

Figure 1.3 Stress-strain graph for ABS polymer, Novodur PK grade. (Courtesy of Bayer UK Ltd)

relationship for plastics depends on the rate at which the strain is applied, unlike metals, where the strain rate is not usually a significant factor. Some typical values of modulus of elasticity and tensile strength for plastics are:

Material *(room temp.)*	*Modulus of elasticity* /GN m^{-2} *or* kN mm^{-2}	*Tensile strength* /MN m^{-2} *or* kN mm^{-2}
Thermoplastics		
Polyvinyl chloride (PVC)	2.5–4.0	35–60
Acrylonitrile-butadiene-styrene (ABS)	2.0–3.0	25–50
Polycarbonate (PC)	2.0–3.0	55–65
Polytetrafluoroethylene (PTFE)	0.3–0.6	15–35
Cellulose acetate (CA)	0.5–2.8	13–62
Thermosets		
Phenol formaldehyde (PH)	5.2–6.0	50–55
Polyester (unsaturated) (UP)	2.0–4.4	40–90

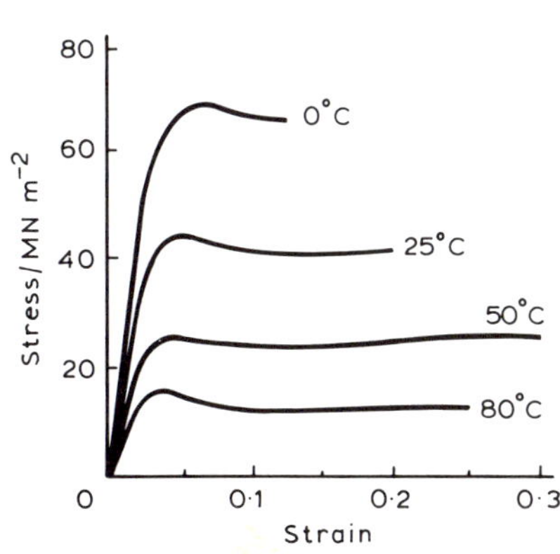

Figure 1.4 Stress-strain graphs for cellulose acetate

The stress/strain properties of plastics, change quite significantly when the temperature changes (*Figure 1.4*). Both the modulus of elasticity and the tensile strength decrease with an increase in temperature.

The stress/strain graph of a plastic can be changed quite significantly if during the processing of the plastic other materials, such as glass fibres, are introduced. *Figure 1.5* shows the change produced in the stress/strain graph of polycarbonate when glass fibres are introduced. The glass fibre material has a higher modulus of elasticity and a higher tensile strength.

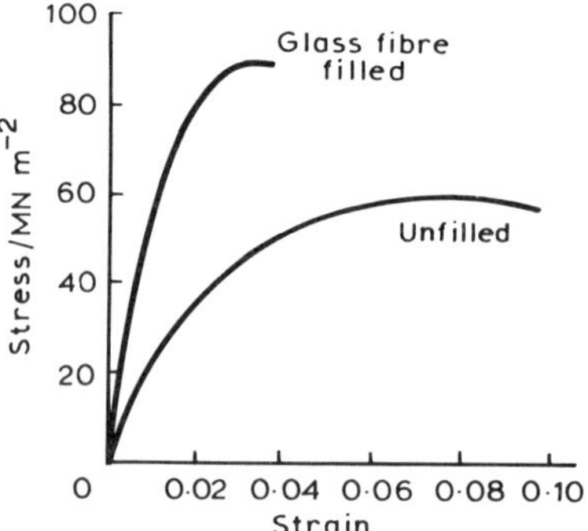

Figure 1.5 Stress-strain graphs for polycarbonate (Makrolon), with and without glass fibre filling. (Courtesy of Bayer UK Ltd)

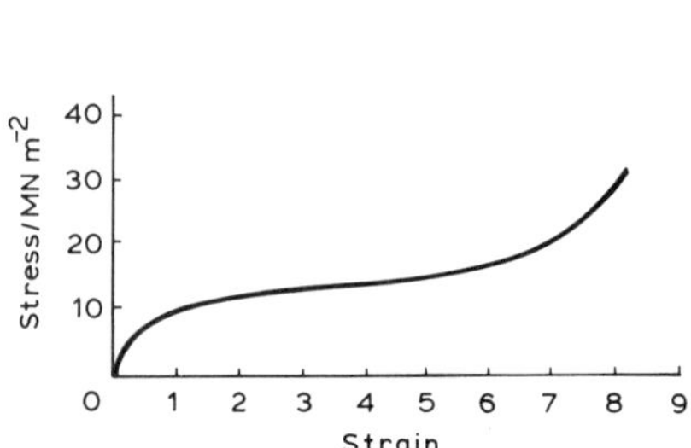

Figure 1.6 Stress-strain graph for rubber

Elastomers are a group of materials that produce very large strains (*Figure 1.6*). Some typical data for elastomers are:

Material	*Tensile strength* /MN m^{-2} *or* N mm^{-2}	*Maximum elongation (%)*
Natural rubber	30	700
Neoprene	28	600
Nitrile	28	550
Silicone	10	700

As will be seen from the form of the stress/strain graph in *Figure 1.6,* the modulus of elasticity is not so useful a quantity for elastomers, as such a modulus can refer to only a very small portion of the stress/strain graph. A typical modulus would be about 30 MN m^{-2}.

HARDNESS

The *hardness* of a material can be defined in terms of the ability of the material to resist scratching or indentation. There is no absolute hardness scale and hardness values are expressed generally in terms of a scale associated with a particular test. In the *Brinell hardness test,* a hardened steel ball is pressed into the surface of the material and the spherical area of the indentation (*Figure 1.7*) determined for the load used. With the magnitude of the force used, this gives a Brinell hardness number (HB). The larger this number the harder the material.

In the *Vickers hardness test* the indenter used is a specially shaped diamond and again the area of the indentation is measured. The Vickers hardness number (HV) is obtained from the area and the force. In the *Rockwell hardness test,* the indenter is a standard diamond cone or hardened steel ball and the permanent increase in

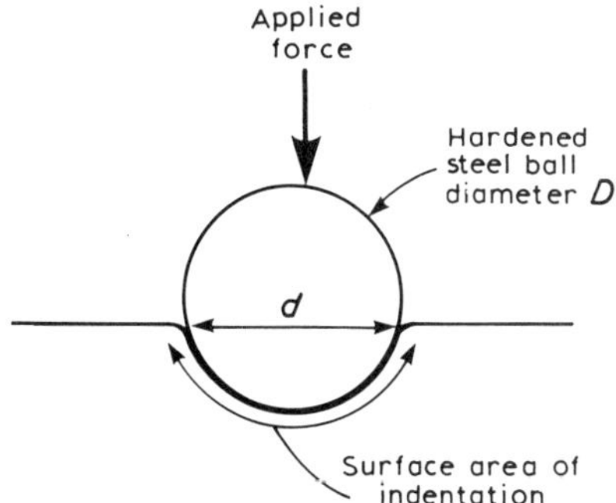

Figure 1.7 The Brinell hardness test. Surface area of indentation is $\pi D/2\ [D-\sqrt{(D^2-d^2)}]$

depth of penetration as a result of a change in the load applied to the indenter is measured. The Rockwell hardness (HR) is given by $(E-e)$, where E is a constant determined by the type of indenter used and e is the depth of penetration. The *Shore scleroscope hardness test* measures the dynamic hardness, not indentation hardness as the others do. Measurement is made of the height of rebound of the indenter after its fall from a fixed height.

The measurement techniques used for the hardness of plastics are similar to those described above for metals. Some typical hardness values for metals and plastics are:

Material	*Hardness (HV)*
Cast iron	140–240
Stainless steel	170
Brass	100–160
ABS	6–8
Polycarbonate	10–20
PVC	9

As can be seen, plastics are much softer than metals.

IMPACT TESTING

With metals, there are two principal forms of impact testing, the *Charpy test* and the *Izod test*. In both tests a heavy pendulum swings through an arc and strikes a notched test piece (*Figure 1.8*). After breaking the sample, the pendulum continues its swing but as some

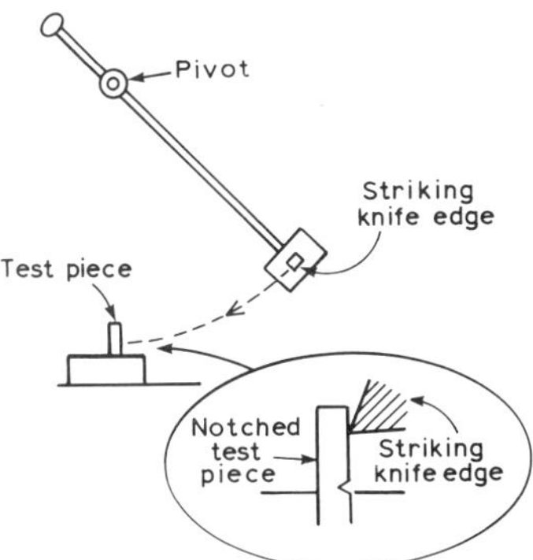

Figure 1.8 The Izod test

energy is used to break the sample, the pendulum does not swing back to the same height as that from which it started. The difference in heights is a measure of the energy absorbed in the breaking. This energy is quoted for standard size samples. The difference between the two tests lies in the mounting of the sample and thus the way in which the pendulum strikes the blow.

Both the Izod and Charpy tests are used with both metals and plastics. In addition, tests are conducted with plastics in which a guided or unguided weight is allowed to fall on a sample of the plastic resting on an annular support.

A brittle material absorbs little energy in breaking, but a ductile material absorbs a greater amount. The term toughness is often used to describe the ability of a material to withstand shock loads, the tougher the material the more it is able to withstand such loads without breaking. The impact tests can be considered as an indicator of toughness. Brittle materials are thus, generally, not very tough.

COMPARISON OF THE PROPERTIES OF DIFFERENT TYPES OF ENGINEERING MATERIALS

Engineering materials can be grouped into four main categories– metals, polymers, ceramics and composites. Polymers include such materials as plastics and elastomers; ceramics include cements, glasses, refractories and abrasives; composites include reinforced plastics, reinforced metals and reinforced ceramics. Reinforced concrete is a composite. These are some of the general characteristics of the various categories:

Property	*Metal*	*Polymer*	*Ceramic*	*Composite**
Relative density	2–16	1–2	2–17	2–3
Melting point/°C	200–3500	70–200	2000–4000	100–200
Tensile strength/MN m^{-2}	100–2500	30–300	10–400	50–1400
Tensile modulus/GN m^{-2}	40–400	0.7–3.5	150–450	10–200
Thermal conductivity	High	Low	Medium	Low
Electrical conductivity	High (conductivity)	Low (insulator)	Low (insulator)	Low (insulator)

* The data refers essentially to reinforced polymers.

Table 1.1

Typical mechanical properties of materials
The following data on the different materials is taken from manufacturers' data sheets.

(a) *Grey iron castings*
(Courtesy of BCIRA)

Tensile strength	180 N/mm^{-2}	
0.01% proof stress	50 N/mm^{-2}	
0.1% proof stress	117 N/mm^{-2}	
Phosphorus (%)	<0.4	0.4–1.0
Total strain at failure (%)	0.7	0.5
Elastic strain at failure (%)	0.17	0.17
Total-minus-elastic strain (%) at failure	0.54	0.33
Compressive strength	672 N/mm^{-2}	
0.01% proof stress	100 N/mm^{-2}	
0.1% proof stress	234 N/mm^{-2}	
Modulus of elasticity		
Tension	109 GN/m^{-2}	
Compression	109 GN/m^{-2}	

Hardness With <0.4% phosphorus 150–183 HB

(b) *Stainless steel*
(Courtesy of Darwins Alloy Castings Ltd)

Group	*Designation*	C	Mn	Si	Cr	Ni	S	P	Fe
Martensitic	S61	0.15 (max)	0.70	0.75	12.50	1.0 (max)	0.045 (max)	0.045 (max)	Balance
	S62	0.16	0.70	0.75	12.50	1.0 (max)	0.045 (max)	0.045 (max)	Balance
	DSC	0.35	0.70	0.75	13.50	1.0 (max)	0.045 (max)	0.045 (max)	Balance

Designation	*Yield point* /N mm^{-2}	*Tensile strength* /N mm^{-2}	*Brinell hardness*	*Modulus of elasticity in tension*/GN m^{-2}
S61	508.0	664.0	193*	200
S62	620.0	772.0	240*	200
DSC	696.0	832.0	279*	200

* Hardened and tempered

(c) *Brass*
(Courtesy of Delta Extruded Metals Co Ltd)
Nominal compositions (%)

	T1	T2	T12	TR1
Copper	57	57.5	57	61.5
Lead	3	2.75	3.9	2.25
Zinc	Rem	Rem	Rem	Rem

Delta alloy	*Condition*	*0.2% proof stress*/MN m^{-2}	*Tensile strength* /MN m^{-2}	*Hardness* (HV)	*Modulus of elasticity* /GN m^{-2}
T1	As extruded	110–140	370–420	90–110	100
	Drawn 6–80 mm	200–260	430–530	140–160	100
	Drawn >80 mm	170–230	380–460	130–150	100
T2	As extruded	110–140	370–420	90–110	100
	Drawn 6–80 mm	200–260	430–530	140–160	100
	Drawn >80 mm	170–230	380–460	130–150	100
T12	As extruded 10–25 mm	110–140	430–460	90–110	100
	Drawn 6 mm	200–260	460–530	140–160	100
TR1	Extruded	140	330	70–80	100
	Drawn 6–50 mm	220–290	370–450	110–130	100
	Drawn >50 mm	190–260	340–420	100–120	100
	Hard	430–480	510–570	150–160	100

(d) *Polycarbonate*
(Courtesy of Bayer UK Ltd)

	Makrolon 2400	Makrolon 8030
Tensile strength/MN m^{-2}	>65	70
Elongation at break (%)	>110	3.5
Yield stress/MN m^{-2}	>55	75
Young's modulus/GN m^{-2}	2.3	5.5

(e) ABS polymer
(Courtesy of Bayer UK Ltd)
Typical characteristics of Novodur, extrusion types.

	PME	PHE
Yield stress/MN m^{-2}	56	45
Elongation (%)	≈2.5	≈3
Young's modulus/GN m^{-2}	2.53	2.34

(f) *PTFE*
(Courtesy of Fothergill & Harvey Ltd)
Tygadure PTFE, moulded and machined components.

Tensile strength/kN m^{-2}		⊁9650–24130
Elongation (%)		⊁100–300
Young's modulus/MN m^{-2}	Compression	482–620
	Tension	275–413
Impact strength (Izod)/J		3.4–5.4

EXPLAINING THE ELASTIC AND PLASTIC BEHAVIOUR OF METALS

A simple theory to explain the elastic and plastic behaviour of metals is the *'block slip' theory,* where a metal is considered to be made of blocks of atoms which can move relative to each other. When a stress is applied to the metal, blocks of atoms become displaced (*Figure 1.9*). When the yield stress is reached there is a movement of large blocks of atoms as they slip past each other, the plane along which this movement occurs being called a *slip plane.* While this theory appears to give a plausible explanation of elastic and plastic behaviour there is one big disadvantage–calculations of the stress needed to displace all the atoms in one plane relative to those in the next plane by at least the 'width' of one atom, indicate a stress value considerably greater than the real results given by experiments. Real metals are not as strong as the theoretical model.

The 'block slip' model has atoms perfectly arranged in an orderly manner within the metal. If, however, we consider the arrangement to be imperfect then permanent deformations can be produced with much less stress. When you have a large carpet which is perfectly flat on the floor, it requires quite an effort to slide the entire carpet and make it move across the floor. But if there is a ruck in the carpet (*Figure 1.10*), then the carpet can be slid over the floor by pushing

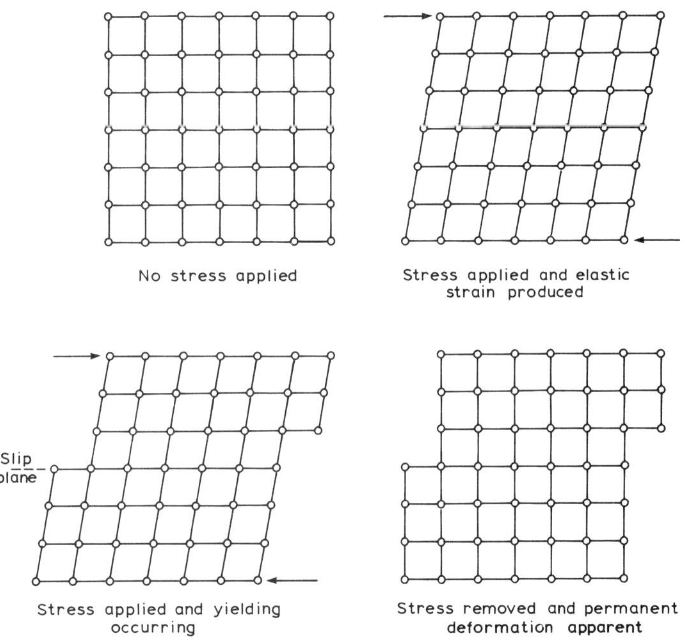

Figure 1.9 'Block-slip' model showing plastic behaviour of metals under stress

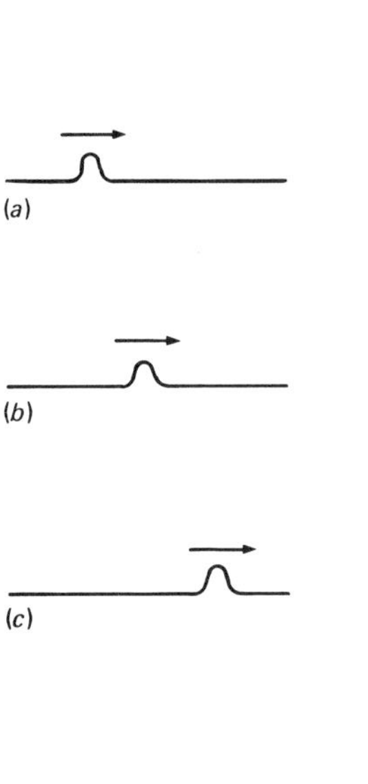

Figure 1.10 Movement of a ruck across a carpet

the ruck along a bit at a time and considerably less effort is required. This is the type of movement which is considered to take place within a metal, the 'ruck' in the crystal being a *dislocation* of atoms due to imperfect packing of the atoms within the metal. *Figure 1.11* shows the type of arrangement of atoms that is considered to occur with what is called an *edge dislocation. Figure 1.12* illustrates the movement that occurs when stress is applied and permanent

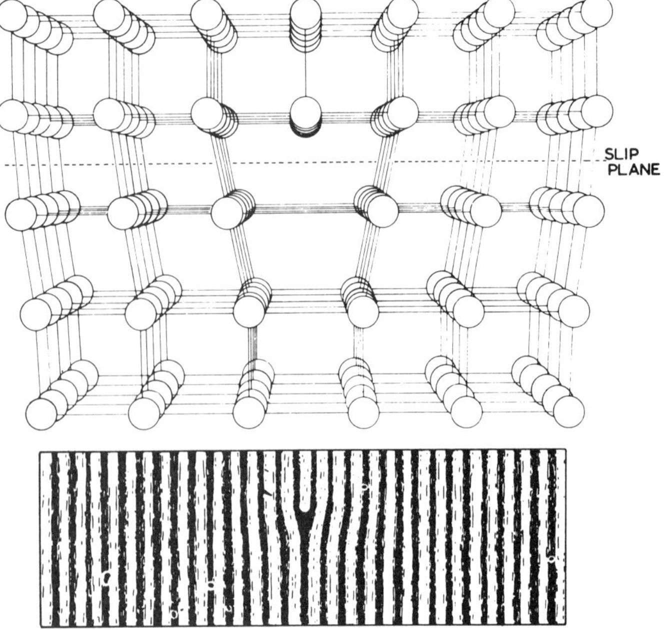

Figure 1.11 A 'ball-and-wire' model of a simple edge dislocation. (From Higgins, R. A., *Properties of Engineering Materials,* Hodder and Stoughton)

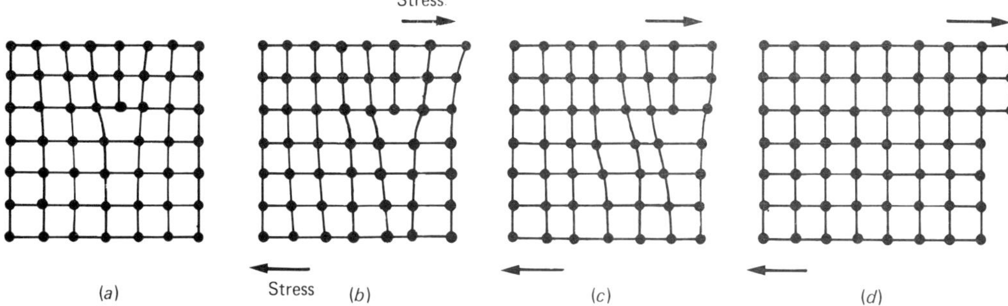

Figure 1.12 Movement of a dislocation through an atomic array under the action of stress

deformation occurs. The dislocation moves through the array of atoms without wholesale movement of planes of atoms past each other; it is a bit-by-bit process like the ruck in the carpet.

Figure 1.13 shows the form of a *screw dislocation* and its movement through the array of atoms under the action of stress.

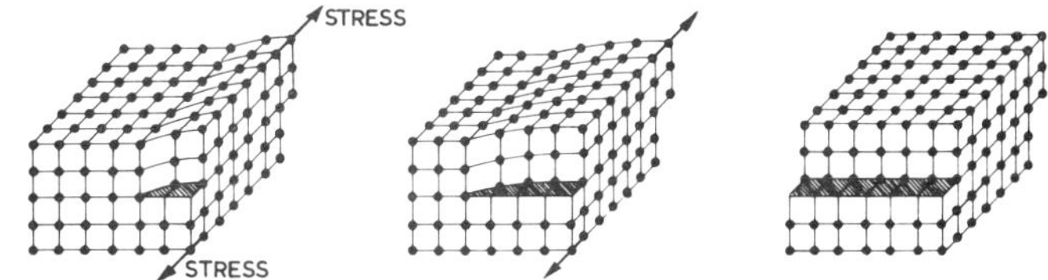

Figure 1.13 Principle of the screw dislocation. (From Higgins, R. A., *Properties of Engineering Materials,* Hodder and Stoughton)

With an edge dislocation the line of dislocation is at right-angles to the slip plane; with a screw dislocation the line of dislocation is parallel to the slip plane. In practice, dislocations are often neither straight lines at right-angles to, or parallel to, the slip plane but curved lines, which can however be considered to be a combination of edge and screw dislocations.

What happens when two dislocations come close to each other during their movement through a metal? As *Figure 1.14* shows, the atoms on one side of the slip plane are in compression and on the other side in tension. When two dislocations come together the regions of compression can impinge on each other and so hinder the movement of the dislocations. If the movement is such as to bring the compression region against the tension region of another

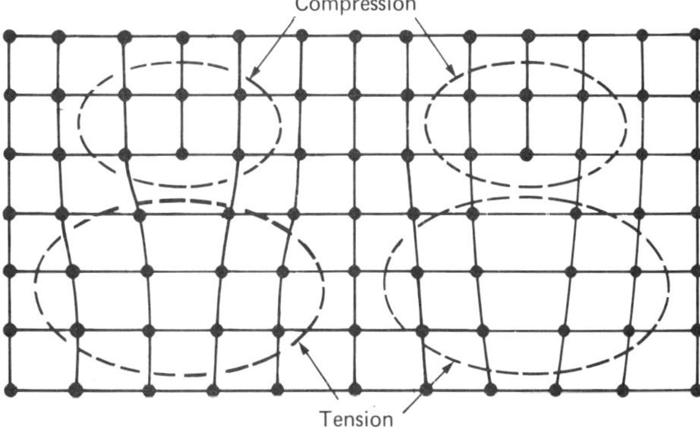

Figure 1.14 Two dislocations in close proximity

dislocation then it is possible for the two dislocations to annihilate each other. In general, the more dislocations a metal has, the more the dislocations get in the way of each other and so the more difficult it is for the dislocations to move through the metal. More stress is needed to cause yielding.

The movement of dislocations through a metal is also hindered by the grain boundaries. The more grain boundaries there are in a metal the more difficult it is to produce yielding of that metal. More grain boundaries occur when the grain size in a metal is small.

The movement of dislocations is hindered by anything that destroys the continuity of the atomic array. The presence of 'foreign' atoms can distort the atomic array of a metal and so hinder the movement of dislocations.

Thus possible ways of increasing the yield stress of a metal are by:

1. Increasing the number of dislocations.
2. Reducing the grain size.
3. Introducing 'foreign' atoms.

Figure 1.15 shows how the dislocation density in niobium increases under increasing plastic strain. If a material is plastically deformed a number of times, i.e. work hardened, the yield stress increases–the dislocation density has increased. The grain size in metals increases if it is heated to about 0.4 times its melting temperature in degrees kelvin. This gives a reduction in yield stress and more easy working conditions for the metal. Hot working of metals requires less energy than cold working, which can be explained by the increase in grain size making it easier for dislocation movement to occur. Alloying is the introduction of 'foreign' atoms into the atomic array of a metal and thus can be expected to change the behaviour of the metal.

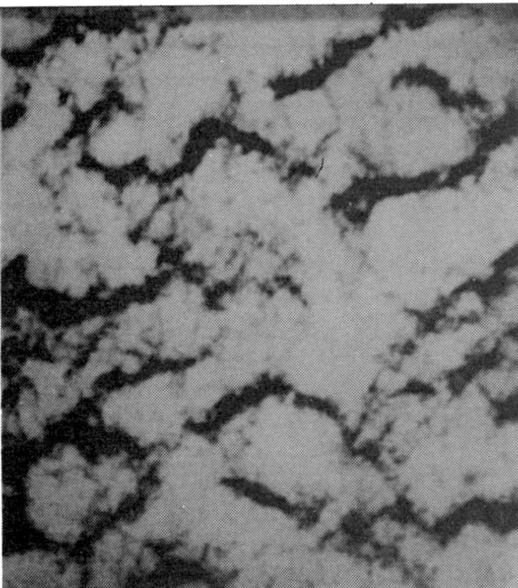

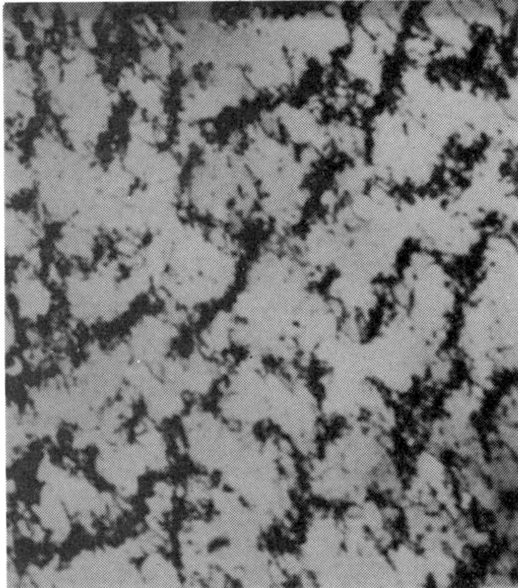

Figure 1.15 Niobium under increasing plastic strain. (From Harris, B. and Bunsell, A. R., *The Structure and Properties of Engineering Materials,* Longmans)

EXPLAINING THE ELASTIC AND PLASTIC BEHAVIOUR OF PLASTICS

Plastics differ from metals in being made up of groups of long molecular chains rather than orderly arrays of atoms. A chain may include a thousand or more atoms. Strong bonds exist between the

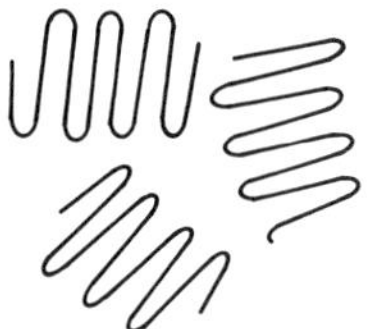

Figure 1.16 Random orientation of coiled molecular chains in an unstressed plastic

atoms in the chains and only, generally, weak bonds between neighbouring chains. In the unstressed plastic, these chains often assume a coiled state, each coiled chain being however randomly orientated (*Figure 1.16*).

Figure 1.17 shows the typical form of stress/strain graph for a thermoplastic material. When stress is applied, the material reasonably uniformly increases in length, the strain being about proportional to the stress. At the yield stress the material begins to 'neck', which is a considerable reduction in the width of the pulled sample. The necking starts at one part of the sample and then extends progressively along its full length. This change is irreversible. When further stress is applied, the material extends more easily than in the early part of the stress/strain graph.

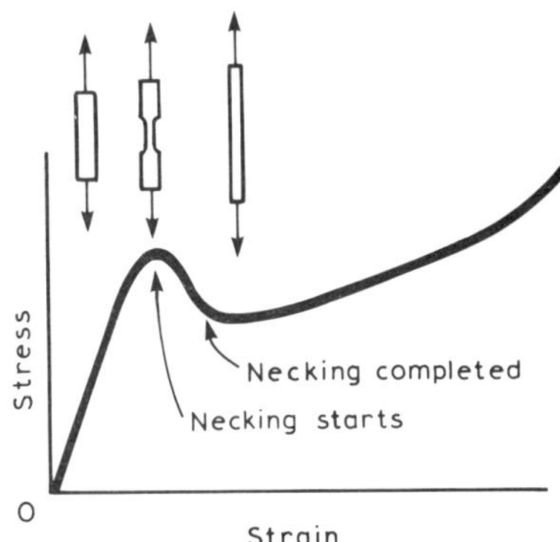

Figure 1.17 Typical stress-strain graph for a thermoplastic

The application of stress to the plastic causes the molecular chains, which were initially randomly orientated, to become orientated in one particular direction. The onset of necking marks the stress at which part of the sample, the necked part, has molecular chains all orientated in one direction. When the necking process is completed all the molecular chains have become orientated in one direction (*Figure 1.18*). Further application of stress results in the chains beginning to uncoil. During this process, a small change in stress can cause quite a large change in extension. Eventually all the chains become uncoiled (*Figure 1.19*) and then further stress causes slippage of these now linear chains past each other.

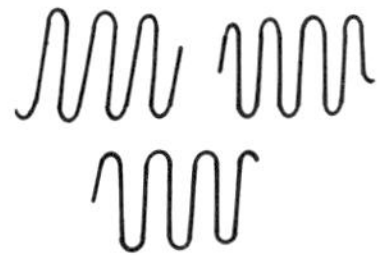

Figure 1.18 Coiled chains all with same orientation when necking completed

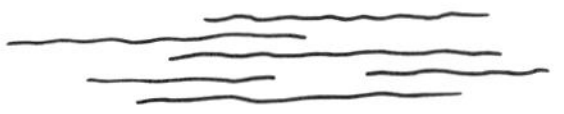

Figure 1.19 Uncoiled chains

In a thermosetting plastic, when the material is in its formed state, there are bonds between the chains in the unstressed state. These bonds prevent the molecular chains straightening out and so a thermosetting material has a different form of stress/strain graph from that of a thermoplastic material.

The above account is much simplified and generalised. More detailed treatment is given in the earlier book in the series, *Materials Technology for Technicians 3.*

PROBLEMS

(1) Use the data given for grey iron castings in Table 1.1a to sketch the form of the stress/strain graph in both tension and compression.

(2) From the stress/strain graph for cast iron (*Figure 1.1b*), determine (a) the modulus of elasticity and (b) the tensile strength of the sample.

(3) From the stress/strain graph for ABS (*Figure 1.3*) determine (a) the modulus of elasticity and (b) the tensile strength of the sample.

(4) Sketch the forms that would be taken by the stress/strain graphs for (a) strong brittle materials, (b) strong ductile materials, (c) weak ductile materials.

(5) Why do some materials have a yield stress quoted whereas others have a proof stress quoted?

(6) For the data given in Table 1.1c for brass: (a) Calculate the maximum force that can be experienced by a sample of T1 brass which has a cross-sectional area of 50 mm^2. (b) Calculate the force needed to reach the 0.2 per cent yield stress for the above T1 sample.

(7) How do the mechanical properties of iron in Table 1.1a and the stainless steel in Table 1.1b differ?

(8) Use the data for polycarbonate in Table 1.1d to sketch the form of the stress/strain graph for Makrolon 2400 and 8030. Describe in words how you would expect the polycarbonate to behave when in service.

(9) A steel is said to have a Charpy V-notch impact value for 10×10 mm specimens at 0°C of 27 J. Explain the significance of this data.

(10) The effect of cold rolling on the properties of aluminium is to make the material harder, e.g. a change from 20 HV in the annealed state to 155 HV in the fully work hardened state. Explain on the basis of dislocation movements why this change in hardness occurs.

(11) Explain, in terms of dislocation movements, why changes in grain size affect the mechanical properties of metals.

(12) Describe the internal changes taking place within a thermoplastic material when, under the action of stress, 'necking' results.

(13) Draw up a table giving comparisons of the properties of the materials, evidenced by the behaviour of tin cans, plastic bottles and ceramic cups.

(14) Investigate the behaviour of metal wire and plastic coat hangers when subject to loads and hence compare the properties of the materials used for the hangers.

(15) Compare the behaviour of metal and plastic spoons when used and hence compare the properties of the materials used for the spoons.

2 Fracture

Objectives: At the end of this chapter you should be able to:
Distinguish between brittle and ductile fractures.
Describe the factors affecting fracture.
Explain the terms strain energy and fracture energy.
Explain brittle failure in terms of Griffith's crack theory.
Interpret impact testing data.

THE PRICE OF A FRACTURE

On Monday 27 December 1965 at about 1.45 p.m., an off-shore drilling rig for North Sea gas called the *Sea Gem* collapsed and sank. Of the 32 men on the rig at the time 19 died.

The rig consisted of a rectangular pontoon with ten steel tubular legs. Each of the legs could be lowered to the sea bed and then the pontoon jacked up on the legs to provide a platform clear of the water. The drilling derrick, operating gear and crew accommodation was on the pontoon. On that Monday, drilling operations having been completed, the crew were preparing to jack the pontoon back down from its fixed platform position. The weight of the pontoon was transferred to the legs by means of tie bars. One or more of the tie bars failed.

A tribunal enquiring into the cause of the accident found that 'the disaster was initiated by brittle fracture of the tie bars at below the yield stress'.

TYPES OF FRACTURE

When a ductile material has a gradually increasing stress applied to it it behaves elastically up to a limiting stress and then beyond that stress plastic deformation occurs. In the case of a tensile stress this deformation takes the form of necking (*Figure 2.1a*). As the stress is increased, the cross-sectional area of the material becomes considerably reduced until at some stress failure occurs. The fracture shows a typical cone and cup formation, which results because under the action of the increasing stress small internal cracks form which

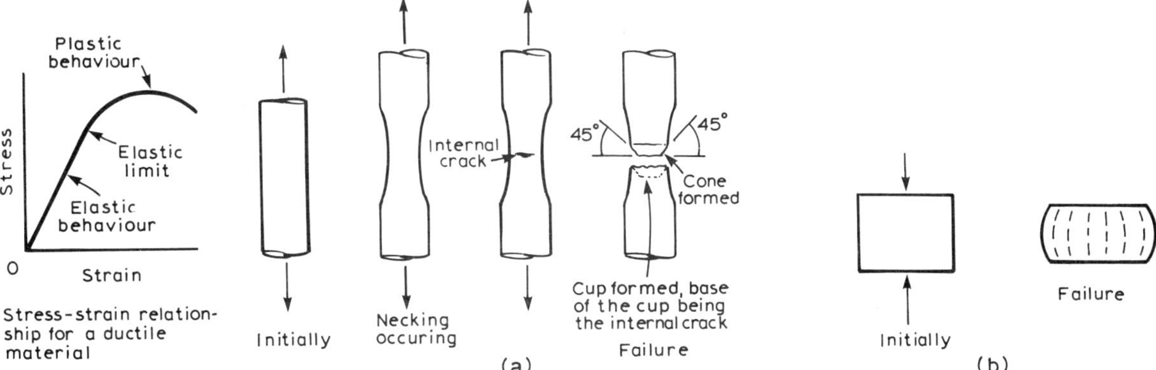

Figure 2.1 Ductile failure (a) Tensile (b) Compressive

gradually grow in size until there is an internal, almost horizontal, crack. The final fracture occurs when the material shears at an angle of 45° to the axis of the direct stress. This type of failure is known as a *ductile fracture.*

Materials can fail in a ductile manner in compression. Such failures show a characteristic bulge and a series of axial cracks around the edge of the material (*Figure 2.1b*).

Another form of failure is known as *brittle fracture.* If you drop a china cup and it breaks it is possible to pick up the pieces and stick them together again and have something which still looks like a cup. The china cup has failed by a brittle fracture. If you had dropped a tin mug then it might have deformed and show a dent. If you could have broken the tin mug it would not have been possible to stick the pieces together and have something that looked like the original tin mug. If somebody drives a car into a wall the metal car wings are most likely to show a ductile type failure like the tin mug. With a brittle failure the material factures before any significant plastic deformation has occurred.

Figure 2.2a shows the possible forms of a brittle tensile failure. The surfaces of the fractured material appears bright and granular due to the reflection of light from individual crystal surfaces. *Figure 2.2b* shows the possible forms of a brittle compressive failure.

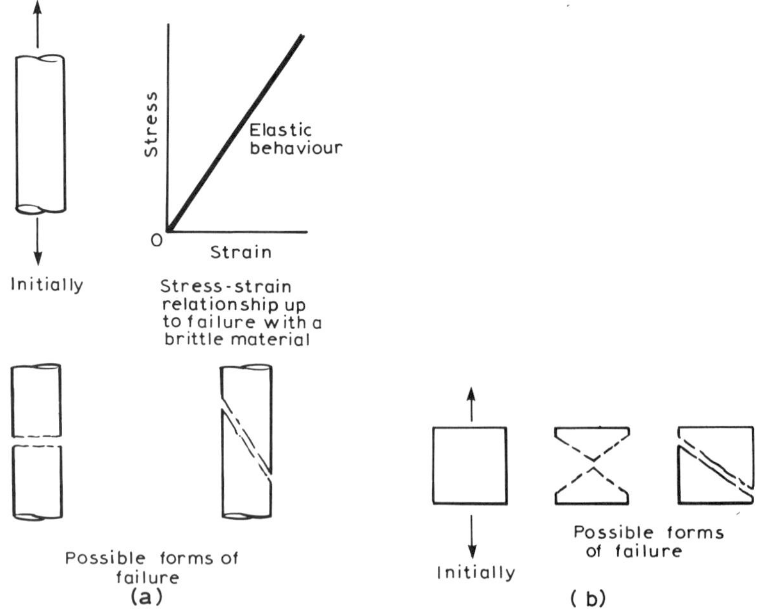

Figure 2.2 Brittle failure (a) Tensile (b) Compressive

FACTORS AFFECTING FRACTURE

If you want to break a piece of material, one way you can adopt is to make a small notch in the surface of the material and then apply a force. The presence of a notch or any sudden change in section of a piece of material can very significantly change the stress at which fracture occurs. The notch or sudden change in section produces what are called *stress concentrations.* They disturb the assumed stress distribution and produce local concentrations of stress.

The amount by which the stress is raised depends on the depth of the notch, or change in section, and the radius of the tip of the notch. The greater the depth of the notch and/or the smaller the radius of the tip of the notch, the greater the amount by which the stress is raised.

A crack in a brittle material will have a quite pointed tip and hence a small radius. Such a crack thus produces a large increase in stress at its tip. One way of arresting the progress of such a crack is to drill a hole at the end of the crack to increase its radius and so reduce the stress concentration.

An approximate relationship that has been derived for the stress at the end of a notch is

$$\text{Stress at end of notch} = \text{applied stress} \times [1 + 2\sqrt{(L/r)}]$$

where L is the length of the notch and r the radius, of the tip of the notch. The increase in stress due to the notch is thus

$$\text{Increase in stress} = 2\sqrt{(L/r)}$$

A crack in a ductile material is less likely to lead to failure than in a brittle material because a high stress concentration at the end of a notch leads to plastic flow and so an increase in the radius of the tip of the notch. The result is a decrease in the stress concentration.

Another factor which can affect the behaviour of a material is the speed of loading. A sharp blow to the material may lead to a fracture where the same stress applied more slowly would not. With a very high rate of application of stress there may be insufficient time for plastic deformation of the material to occur and so what was under normal conditions a ductile material behaves as though it were brittle. If a material has a notch or change in section and is subject to a sudden impact, the dual effects of the stress concentration due to the notch and the material behaving as though brittle can result in failure.

The Charpy and Izod tests referred to in Chapter 1 give a measure of the behaviour of a notched sample of material when subject to a sudden impact load. The results are expressed in terms of the energy needed to break a standard size test specimen; the smaller the energy needed to break the specimen, the easier it will be for failure to occur in service. The smaller energies are associated with materials which are termed brittle; ductile materials need higher energies for fracture to occur.

The temperature of a material when it is subject to stress can affect its behaviour, many metals which are ductile at high temperatures being brittle at low temperatures. *Figure 2.3* shows how the impact

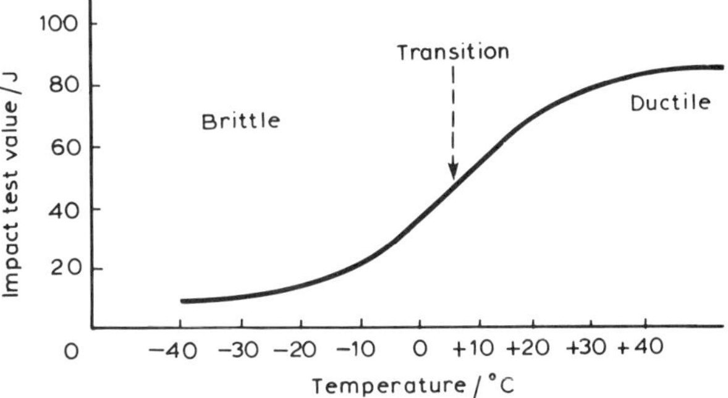

Figure 2.3 Ductile/brittle transition

test results for a steel change with temperature. At room temperature and above the steel behaves as a ductile material; below 0°C it behaves as a brittle material. The *transition* temperature at which the change from ductile to brittle behaviour occurs is thus of importance in determining how a material will behave in service.

The transition temperature with a steel is affected by the alloying elements in the steel. Manganese and nickel reduce the transition temperature. Thus for low-temperature work a steel with these alloying elements should be preferred. Carbon, nitrogen and phosphorus increase the transistion temperature.

The failure of the *Sea Gem* drilling rig referred to at the beginning of this chapter was due to a brittle fracture. On the day of the failure the temperature were low – about 3°C. A Charpy test carried out on a sample of the steel under the same conditions showed an impact test value of only about 10 to 30 J. The part that failed had a change in section with a relatively small fillet radius, about 5 mm and tests showed that this would result in a stress concentration which would give an increase in stress by a factor of about 7. All these factors are conducive to the type of failure that actually occurred – a brittle fracture.

STRAIN ENERGY

Energy is needed to extend a piece of material. There is a transfer of energy from some source to the material. *Figure 2.4* shows the linear part of a force/extension graph for a material. To cause the

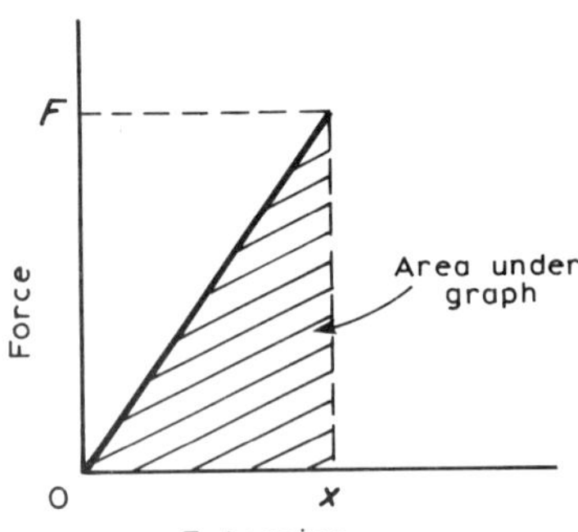

Figure 2.4

material to extend and have an extension x a force is applied which gradually increases from zero to F. The average force is $F/2$ and thus the energy transferred to the material is given by

$$\text{Energy transferred} = \text{force} \times \text{distance}$$
$$= \tfrac{1}{2}Fx$$

This is the area under the graph between zero extension and extension x.

The volume of material being extended is AL, where A is the cross-sectional area and L the length. Hence the energy per unit volume is

$$\frac{\text{energy}}{\text{volume}} = \frac{Fx}{2AL}$$
$$= \tfrac{1}{2}Fx/AL$$

But F/A is the stress and x/L the strain, hence

energy per unit volume $= \frac{1}{2}$ stress $\times$ strain

Units: stress, N/m^2; strain, no units; energy per unit volume, J/m^3 or N/m^2.

If the material is stretched only within its elastic limit, then when the stress is removed, the material springs back to its original dimensions. The energy used to extend the material is surrendered by the material. A stretched material thus has a store of energy, called *strain energy*. For example, a catapult relies on energy being slowly stored in the rubber of the catapult which is suddenly released and given to a projectile.

FRACTURE ENERGY

Energy is needed to stretch a material, and if the stretching continues far enough the material will break. Thus energy is needed to break a material. When a material is broken, new material surfaces are created (*Figure 2.5*). The energy needed to produce unit area of new surface is often called the *fracture energy* or *work of fracture.*

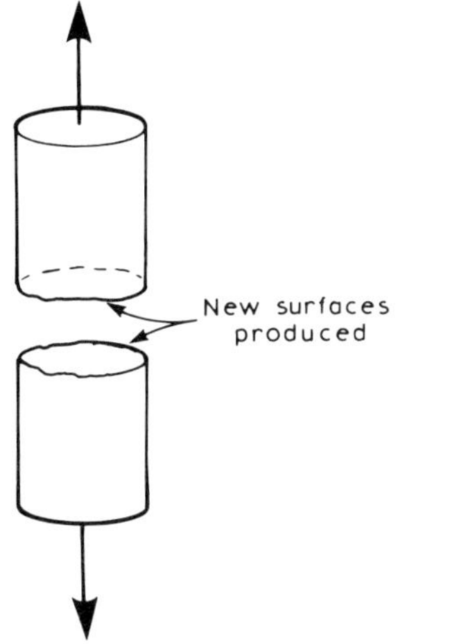

Figure 2.5 New surfaces produced at fracture

Strain energy is related to the area under the stress/strain graph. In the same way, the fracture energy is related to the area under the stress/strain graph – this time the entire area up to the breaking strain (*Figure 2.6*). The bigger this area the bigger the fracture energy. Mild steel has a fracture energy of about 10^5 to 10^6 J m^{-2}; the material used for the china tea cup has a fracture energy of only 1 to 10 J m^{-2}. Tea cups are easier to break than mild steel.

The greater the cross-sectional area of a piece of material the greater the energy needed to break it. Thus, for a piece of mild steel with a fracture energy of 10^5 J/m^{-2} and a cross-sectional area of 0.01 m^2 the energy needed for fracture is $10^5 \times 0.01 = 1000$ J.

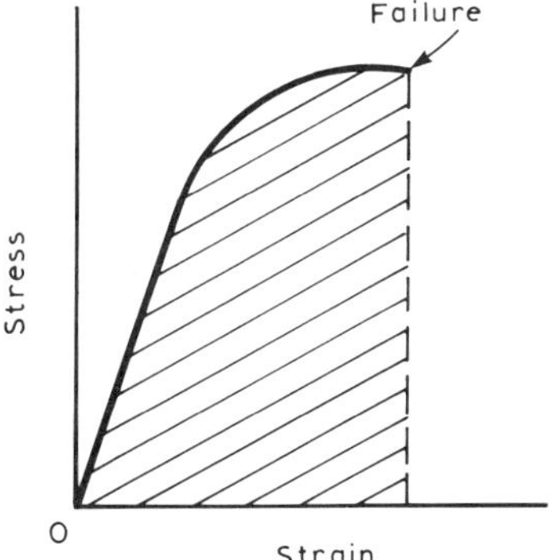

Figure 2.6

GRIFFITH CRACK THEORY

Materials used in any structure or component are likely to have small surface scratches, contain holes and even small cracks. Despite all these, the structure or component may function perfectly satisfactorily under stress. The stress does not cause the scratches, holes or cracks to propagate and so produce failure. A sheet of postage stamps has rows of perforations between the stamps and generally a slight tearing motion starting at one hole will cause a 'crack' to propagate along the row of perforations – that is what the perforations are there for. Why don't the holes, scratches and cracks in a material always propagate when stress is applied?

In 1920, A. A. Griffith advanced the theory that all materials contain small cracks but that at any particular stress there is a *critical crack length* below which propagation of the crack will not occur. The relationship found by Griffith was

Critical crack length

$$c = \frac{1}{\pi} \times \frac{\text{fracture energy}}{\text{strain energy stored per unit volume of material}}$$

The fracture energy is needed to create the surfaces of a crack, i.e. to separate the atoms in the solid so that surfaces are created. Strain energy is however stored in a material, and this is released by a crack. If the energy released when a crack is produced is big enough to create new surfaces, i.e. pull atoms apart, then the crack propagates, but if the energy release is not big enough then the crack does not propagate.

For a material which fails by brittle fracture, the stress is proportional to the strain up to the stress at which fracture occurs. In such a case,

$$\begin{aligned}\text{stored strain energy/unit volume} &= \tfrac{1}{2}\,\text{stress} \times \text{strain}\\ &= \tfrac{1}{2}\,\text{stress} \times (E \times \text{stress})\end{aligned}$$

where E is the modulus of elasticity. Thus the critical crack length equation can be written, for such a mode of failure, as

$$c = 2E \times \text{fracture energy}/\pi\sigma^2$$

where σ is the average tensile stress in the material near the crack, disregarding any effect of stress concentration due to the presence of the crack. Thus for a given material,

$$c \propto 1/\sigma^2$$

The bigger the stress the smaller the critical crack length.

If you stick a pin in the rubber of a deflated balloon all that happens is that a pin hole is produced; the hole does not result in a crack propagating, so the stress is low. If you stick a pin in the rubber of an inflated balloon there will probably be a bang as the crack started by the hole propagates very rapidly through the rubber. In the inflated balloon situation, the rubber is under sufficient stress for the critical crack length to be very small, smaller than the size of the pinhole.

The energy needed to produce a tensile failure of mild steel is about 10^5 to 10^6 J per square metre. The modulus of elasticity is about 200×10^9 N/m^2. Thus if mild steel is to have cracks which do not propagate at, say, a stress of 100×10^6 N/m^2 then the cracks must be smaller than

$$\begin{aligned}c &= \frac{2 \times 200 \times 10^9 \times 10^5}{\pi \times (100 \times 10^6)^2}\\ &= 1.3\ \text{m, to two significant figures.}\end{aligned}$$

The lower value of the energy was chosen for the calculation. If twice the stress were encountered, i.e. 200×10^6 N/m^2, then the critical crack length is only about 0.3 m.

Despite Griffith's theory, small cracks can still propagate. The theory is useful however in determining whether a crack is likely to propagate quickly.

PROBLEMS

(1) Distinguish between the forms of ductile and brittle fractures.

(2) Given a fractured specimen what would you look for in order to determine whether the fracture was a ductile or a brittle fracture?

(3) How does the presence of a notch or an abrupt change in section have an effect on the failure behaviour?

(4) Explain with the aid of sketches of stress/strain graphs why, in general, less energy is needed for a material to fail by brittle fracture than by ductile fracture.

(5) The material of a china cup has a fracture energy of about 1 to 10 J/m^2. The material of a plastic cup has a fracture energy of about 1000 J/m^{-2}. Why are china cups easier to break than plastic cups? The tensile strength of the china cup material is about 150–200 MN/m^{-2}, that of the plastic about 150–600 MN/m^{-2}.

(6) In 1944 in Cleveland, Ohio, a steel tank holding liquefied natural gas fractured and caused 128 deaths and considerable damage. The temperature to which the steel was exposed was very low, of the order of −160°C. What effect might this low temperature have had on the steel? Why would the choice of steel for such an application be important? What types of steel may be suitable?

(7) Describe how, for a steel, the results given by an impact test such as the Charpy test, change with temperature when the steel shows a ductile-brittle transition.

(8) Explain the term critical crack length as used in the Griffith crack theory.

(9) With mild steel having a fracture energy of 100 kJ m^{-2} and a modulus of elasticity of 200 GN m^{-2} what is the critical crack length if the steel is to be subject to a tensile stress of 120 MN m^{-2}?

(10) A high tensile steel with a fracture energy of 10 kJ m^{-2} and a modulus of elasticity of 200 GN m^{-2} contains a crack of length 200 mm. At what stress would the crack be expected to propagate and result in complete failure?

(11) The following data was taken from information published by GKN Steelstock Ltd, Colnbrook, Slough, England. Steel plate to BS 4360:1979.

Grade	*Temperature*/°C	*Charpy V-notch impact test*/J	*Tensile strength*/N mm^{-2}
50C	−5	41	490/620
	−15	27	
50D	−20	41	490/620
	−30	27	
55E	−20	61	550/700
	−30	47	
	−50	27	

(a) What information is given by Charpy V-notch impact tests?

(b) Explain the significance of the data on the use that can be made of the different grade steel plate.

(12) Estimate the fracture energies for mild steel and cast iron from the stress/strain graphs in *Figure 1.1.*

(13) Do plastic spoons exhibit brittle or ductile fracture when broken at room temperature? Does the type of fracture depend on the temperature of the spoon? Try an investigation.

3 Fatigue

Objectives: At the end of this chapter you should be able to:
Explain what is meant by fatigue failure.
Describe the types of fatigue tests used, explaining the terms stress, amplitude, fatigue limit, endurance limit and S/N graphs.
Describe the main factors affecting the fatigue properties of metals.
Describe the fatigue properties of plastics.
Interpret fatigue data.

FATIGUE FAILURE

If you take a stiff piece of metal or plastic, and want to break it, then you will most likely flex the strip back and forth, as in *Figure 3.1.* This is generally an easier way of causing the material to fail then applying a direct pull.

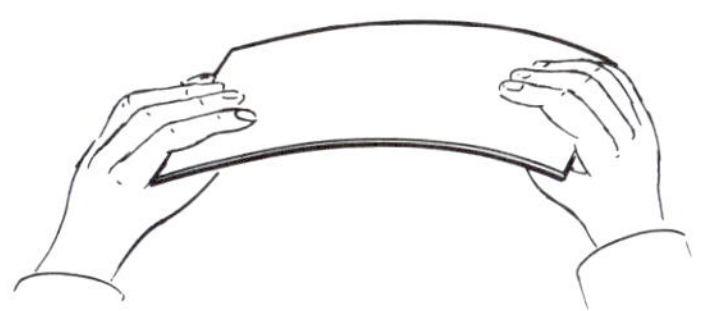

Figure 3.1

In service many components undergo thousands, often millions, of changes of stress. Some are repeatedly stressed and unstressed, while some undergo alternating stresses of compression and tension. For others the stress may just fluctuate about some value. Many materials subject to such conditions fail, even though the maximum stress in any one stress change is less than the fracture stress as determined by a simple tensile test. Such a failure, as a result of repeated stressing, is called a *fatigue failure* (*Figure 3.2*).

The source of the alternating stresses can be due to the conditions of use of a component. Thus, in the case of an aircraft, the changes

Figure 3.2 Fatigue failure of a large shaft. (From John, V. B., *Introduction to Engineering Materials,* Macmillan)

of pressure between the cabin and the outside of the aircraft every time it flies subject the cabin skin to repeated stressing. Components such as a crown wheel and pinion are subject to repeated stressing by the very way in which they are used, while others receive their stressing 'accidently'. Vibration of the component can occur as a result of the transmission of vibration from some machine nearby. Turbine blades may vibrate in use in such a way that they fail by fatigue. It has been said that fatigue causes at least 80 per cent of the failures in modern engineering components.

A fatigue crack often starts at some point of stress concentration. This point of origin of the failure can be seen on the failed material as a smooth, flat, semicircular or elliptical region, often referred to as the nucleus. Surrounding the nucleus is a burnished zone with ribbed markings. This smooth zone is produced by the crack propagating relatively slowly through the material and the resulting fractured surfaces rubbing together during the alternating stressing of the component. When the component has become so weakened by the crack that it is no longer able to carry the load, the final, abrupt fracture occurs, which shows a typically crystalline appearance. *Figure 3.3* shows the various stages in the growth of a fatigue crack failure.

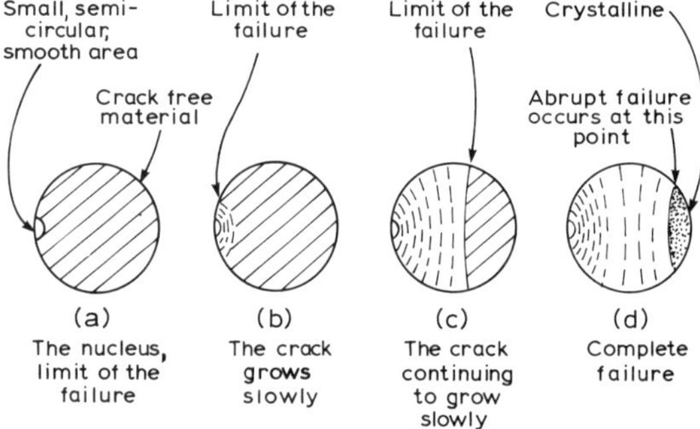

Figure 3.3 Fatigue failure with a metal

FATIGUE TESTS

Fatigue tests can be carried out in a number of ways, the way used being the one needed to simulate the type of stress changes that will occur to the material of a component when in service. There are thus bending-stress machines which bend a test piece of the material alternately one way and then the other (*Figure 3.4a*), and torsional-fatigue machines which twist the test piece alternately one way and then the other (*Figure 3.4b*). Another type of machine can be used to produce alternating tension and compression by direct stressing (*Figure 3.4c*).

The tests can be carried out with stresses which alternate about zero stress (*Figure 3.4d*), apply a repeated stress which varies from zero to some maximum stress (*Figure 3.4e*) or apply a stress which varies about some stress value and does not reach zero at all (*Figure 3.4f*).

In the case of the alternating stress (*Figure 3.4d*), the stress varies between $+S$ and $-S$. The tensile stress is denoted by a positive sign, the compressive stress by a negative sign; the stress range is thus $2S$. The mean stress is zero as the stress alternates equally about the zero stress. With the repeated stress (*Figure 3.4e*), the mean stress is half the stress range. With the fluctuating stress (*Figure 3.4f*) the mean stress is more than half the stress range.

In the fatigue tests, the machine is kept running, alternating the stress, until the specimen fails, the number of cycles of stressing up to failure being recorded by the machine. The test is repeated for the specimen subject to different stress ranges. Such tests enable graphs similar to those in *Figure 3.5* to be plotted. The vertical axis is the *stress amplitude,* half the stress range. For a stress amplitude greater than the value given by the graph line, failure occurs for the number of cycles concerned. These graphs are known as *S/N graphs,* the *S* denoting the stress amplitude and the *N* the number of cycles.

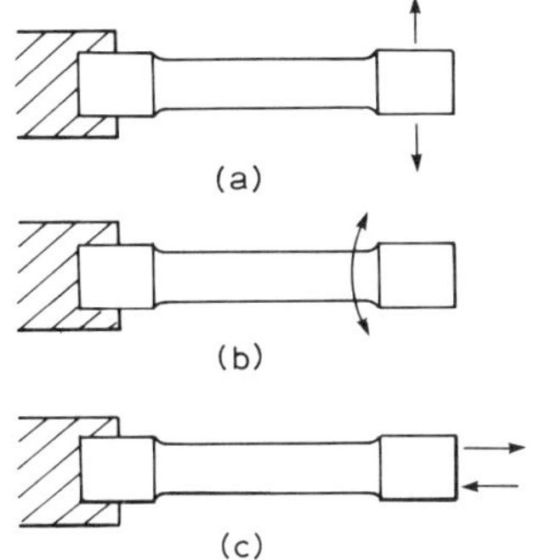

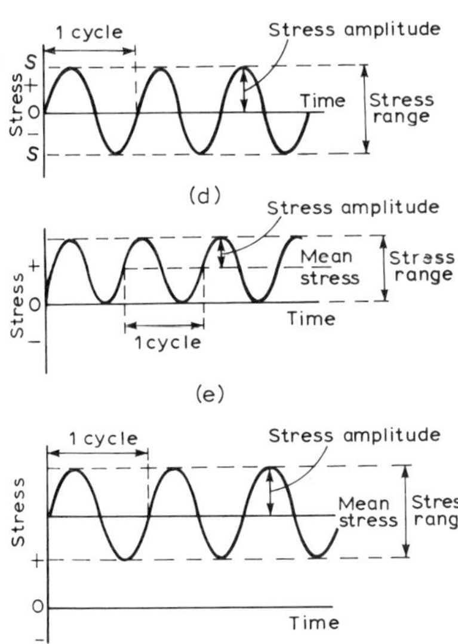

Figure 3.4 Fatigue testing (a) Bending (b) Torsion (c) Direct stress (d) Alternating stress (e) Repeated stress (f) Fluctuating stress

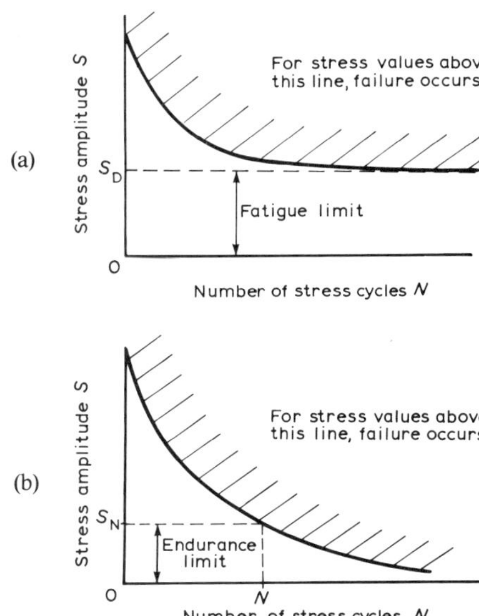

Figure 3.5 Typical *S/N* graphs for (a) A steel (b) A non-ferrous alloy

For the *S/N* graph in *Figure 3.5a* there is a stress amplitude for which the material will endure an indefinite number of stress cycles. The maximum valuc, S_D, being called the *fatigue limit.* For any stress amplitude greater than the fatigue limit, failure will occur if the material undergoes a sufficient number of stress cycles. With the *S/N* graph shown in *Figure 5.3b* there is no stress amplitude at which failure cannot occur; for such materials as *endurance limit* S_N is quoted. This is defined as the maximum stress amplitude which can be sustained for *N* cycles.

The number of reversals that a specimen can sustain before failure occurs depends on the stress amplitude, the bigger the stress amplitude the smaller the number of cycles of stress reversals that can be sustained.

Some typical results for an aluminium alloy specimen are:

Stress amplitude/MN m^{-2}	*Number of cycles before failure*/ $\times 10^6$
185	1
155	5
145	10
120	50
115	100

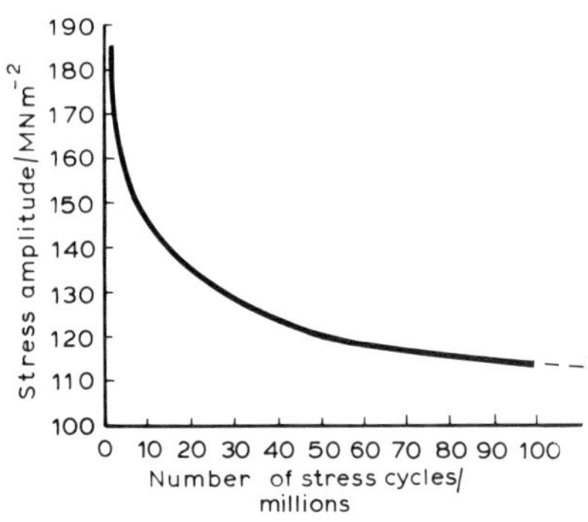

Figure 3.6 *S/N* graph for an aluminium alloy

With a stress amplitude of 185 MN m^{-2}, e.g. a stress alternating from +185 MN m^{-2} to −185 MN m^{-2}, one million cycles are needed before failure occurs. With a smaller stress amplitude of 115 MN m^{-2} one hundred million cycles are needed before failure occurs. *Figure 3.6* shows the *S/N* graph for the above data. Extrapolation of the graph seems to indicate that for a greater number of cycles, failure will occur at even smaller stress amplitudes. There seems to be no stress amplitude for which failure will not occur; the material has no fatigue limit. If a component made of that material had a service life of 100 million stress cycles then we could specify that during the lifetime, failure should not occur for stress amplitudes less than 115 MN m^{-2}. The endurance limit for 100 million cycles is thus 115 MN m^{-2}.

The following are some typical results for a steel, the fatigue tests being bending stress (*Figure 3.7*):

Stress amplitude/MN m^{-2}	*Number of cycles before failure* ($\times 10^6$)
750	0.01
550	0.1
450	1
450	10
450	100

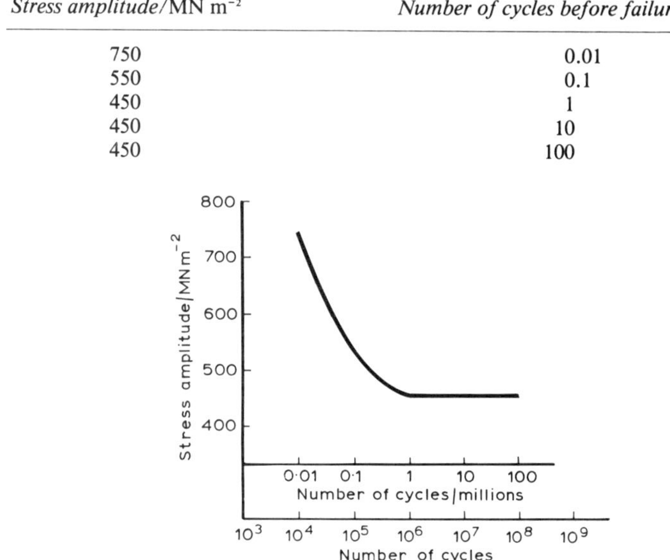

Figure 3.7 *S/N* graph for a steel. Note: number of cycles shown on logarithmic scale

With a stress amplitude of 750 MN m^{-2}, e.g. a stress alternating from +750 NM m^{-2} to −750 MN m^{-2}, 0.01 million cycles or ten thousand cycles are needed before failure occurs. For one million, ten million and one hundred million cycles the stress amplitude for failure is the same, 450 MN m^{-2}. For stress amplitudes below this value the material should not fail, however long the test continues. The fatigue limit is thus 450 MN m^{-2}.

The fatigue limit, or the endurance limit at about 500 million cycles, for metals tends to lie between about a third and a half of the static tensile strength. This applies to most steels, aluminium alloys, brass, nickel and magnesium alloys. For example, a steel with a tensile strength of 420 MN m^{-2} has a fatigue limit of 180 MN m^{-2}, just under half the tensile strength. If used in a situation where it were subject to alternating stresses, such a steel would need to be limited to stress amplitudes below 180 MN m^{-2} if it were not to fail at some time. A magnesium alloy with a tensile strength of 290 MN m^{-2} has an endurance limit of 120 MN m^{-2}, just under half the tensile strength. Such an alloy would need to be limited to stress amplitudes below 120 MN m^{-2} if it were to last to 500 million cycles.

Table 3.1 Typical manufacturers' information concerning grey cast irons (Courtesy of BCIRA)

Fatigue limit

For grey cast irons with tensile strengths of 150-300 N mm^{-2} the fatigue limit is about 45 per cent of the tensile strength. At higher stresses the ratio decreases, and values of 0.425 and 0.38 have been considered typical of irons with tensile strengths of 350 and 400 N mm^{-2}.

Grey cast irons are relatively insensitive to notches in fatigue; normally a notch will not reduce the fatigue limit in an iron with a tensile strength of 150 N mm^{-2}. In an iron with a tensile strength of 400 N mm^{-2} notching has been found to reduce the fatigue limit to about 83 per cent of the unnotched value.

FACTORS AFFECTING THE FATIGUE PROPERTIES OF METALS

The main factors affecting the fatigue properties of a component are:

1. Stress concentrations caused by component design.
2. Corrosion.
3. Residual stresses.
4. Surface finish.
5. Temperature.

Fatigue of a component depends on the stress amplitude attained, the bigger the stress amplitude the fewer the stress cycles needed for failure. Stress concentrations caused by sudden changes in cross-section, keyways, holes or sharp corners can thus more easily lead to a fatigue failure. The presence of a countersunk hole was considered in one case to have lead to a stress concentration which could have led to a fatigue failure. *Figure 3.8* shows the effect on the fatigue properties of a steel of a small hole acting as a stress raiser. With the hole, at every stress amplitude value less cycles are needed to reach failure. There is also a lower fatigue limit with the hole present, 700 MN m^{-2} instead of over 1000 MN m^{-2}.

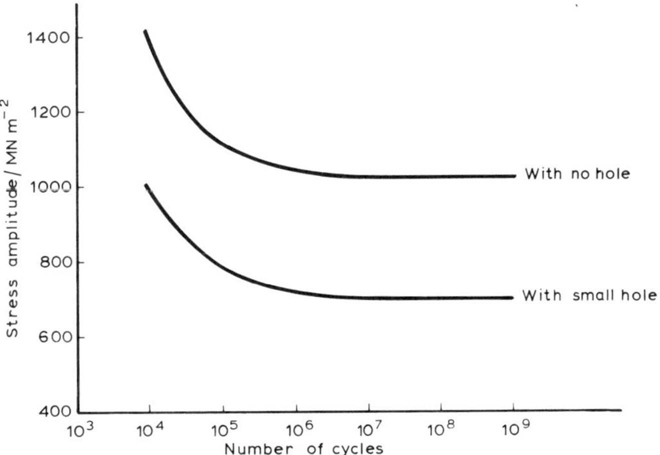

Figure 3.8 *S/N* graph for a steel both with and without small hole acting as stress raiser. Note: number of cycles shown on logarithmic scale

Figure 3.9 shows the effect on the fatigue properties of a steel of exposure to salt solution. The effect of the corrosion resulting from the salt solution attack on the steel is to reduce the number of stress cycles needed to reach failure for every stress amplitude. The non-corroded steel has a fatigue limit of 450 MN m^{-2}, the corroded steel has no fatigue limit. There is thus no stress amplitude below which failure will not occur. The steel can be protected against the corrosion by plating; for example, chromium or zinc plating of the steel can result in the same *S/N* graph as the non-corroded steel even though it is subject to a corrosive atmosphere (see Chapter 5 for a further discussion of corrosion).

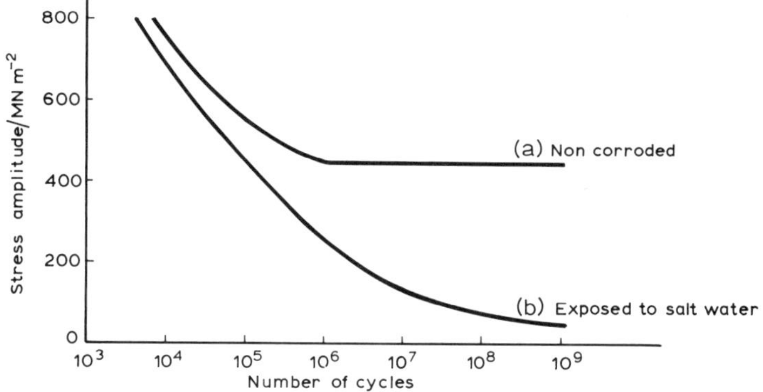

Figure 3.9 *S/N* graph for a steel, (a) with no corrosion and (b) corroded by exposure to salt solution. Note: number of cycles shown on logarithmic scale

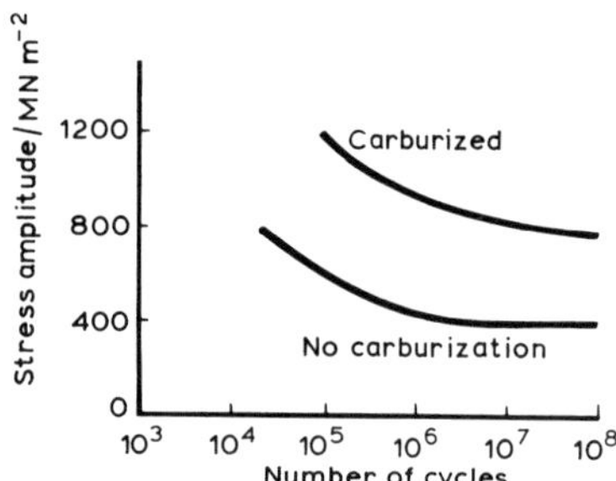

Figure 3.10 *S/N* graph for a steel, showing effect of carburisation. Note: number of cycles shown on logarithmic scale

Residual stresses can be produced by many fabrication and finishing processes. If the stresses produced are such that the surfaces have compressive residual stresses then the fatigue properties are improved, but if tensile residual stresses are produced at the surfaces then poorer fatigue properties result. The case-hardening of steels by carburising results in compressive residual stresses at the surface, hence carburising improves the fatigue properties. *Figure 3.10* shows the effect of carburising a hardened steel. Many machining processes result in the production of surface tensile residual stresses and so result in poorer fatigue properties.

The effect of surface finish on the fatigue properties of a component is very significant. Scratches, dents or even surface identification markings can act as stress raisers and so reduce the fatigue properties. Shot peening a surface produces surface compressive residual stresses and improves the fatigue performance.

An increase in temperature can lead to a reduction in fatigue properties as a consequence of oxidation or corrosion of the metal surface increasing. For example, the nickel-chromium alloy Nimonic 90 undergoes surface degradation at temperatures around 700 to 800°C and there is a poorer fatigue performance as a result. In many instances an increase in temperature does result in a poorer fatigue performance.

THE FATIGUE PROPERTIES OF PLASTICS

Fatigue tests can be carried out on plastics in the same way as with metals. A factor not present with metals is that when a plastic is subject to an alternating stress it becomes significantly warmer. The faster the stress is alternated, i.e., the higher the frequency of the alternating stress, the greater the temperature rise. Under very high frequency alternating stresses the temperature rise may be large enough to melt the plastic. To avoid this, fatigue tests are normally carried out with lower-frequency alternating stresses than is usual with metals. The results of such tests, however, are not entirely valid if the alternating stresses experienced by the plastic component in service are higher than those used for the test.

Figure 3.11 shows an *S/N* graph for a plastic, unplasticised p.v.c. The alternating stresses were applied with a square waveform at a frequency of 0.5 Hz, i.e. a change of stress every 2 s. The graph seems to indicate that there will be no stress amplitude for which failure will not occur; the material thus seems to have no fatigue limit.

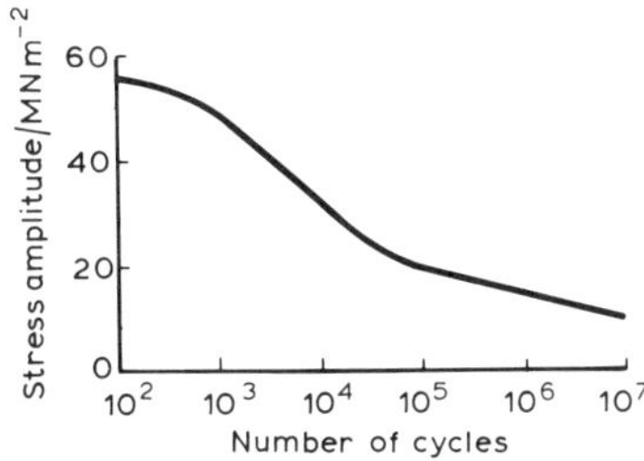

Figure 3.11 *S/N* graph for unplasticised p.v.c., alternating stress being a square waveform at frequency 0.5 Hz. Note: number of cycles shown on logarithmic scale

PROBLEMS

(1) Explain what is meant by fatigue failure.

(2) List the types of test available for the determination of the fatigue properties of specimens.

(3) Describe the various stages in the failure of a component by fatigue.

(4) Explain the terms 'fatigue limit' and 'endurance limit'.

(5) *Figure 3.6* shows the *S/N* graph for an aluminium alloy.

(a) For how many stress cycles could a stress amplitude of 140 MN m^{-2} be sustained before failure occurs?

(b) What would be the maximum stress amplitude that should be applied if the component made of the material is to last for 50 million stress cycles?

(c) The alloy has a tensile strength of 400 MN m^{-2} and a yield stress of 280 MN m^{-2}. What should be the limiting stress when such an alloy is used for static conditions? What should be the limiting stress when the alloy is used for dynamic conditions where the number of cycles is not likely to exceed 10 million?

(6) Explain an *S/N* graph and state the information that can be extracted from the graph.

(7) What is the fatigue limit for the uncarburised steel giving the *S/N* graph in *Figure 3.10*?

(8) What is the endurance limit for the unplasticized p.v.c. at 10^6 cycles that gave the *S/N* graph in *Figure 3.11*?

(9) Plot the *S/N* graph for the nickel-chromium alloy Nimonic 90, which gave the following fatigue test results. Determine from the graph the fatigue limit.

Stress amplitude/MN m^{-2}	*Number of cycles before failure*
750	10^5
480	10^6
350	10^7
320	10^8
320	10^9
320	10^{10}

(10) Plot the *S/N* graph for the plastic (cast acrylic) which gave the following fatigue test results when tested with a square waveform at 0.5 Hz. Why specify this frequency? What is the endurance limit at 10^6 cycles?

Stress amplitude/MN m^{-2}	*Number of cycles before failure*
70	10^2
62	10^3
58	10^4
55	10^5
41	10^6
31	10^7

(11) *Figure 3.12* shows the *S/N* graph for a nickel-based alloy Iconel 718.

(a) What is the fatigue limit?

(b) What is the significance of the constant stress amplitude part of the graph from 10^0 to 10^4 cycles?

(c) The graph is for the material at 600°C. The tensile strength at that temperature is 1000 MN m^{-2}. What can be added to your answer to part (b)?

(12) *Figure 3.13* shows two *S/N* graphs, one for the material in an un-notched state, the other for the material with a notch. Which of the graphs would you expect to represent each condition? Give reasons for your answer.

(13) List factors that contribute to the onset of fatigue failure and those which tend to resist fatigue.

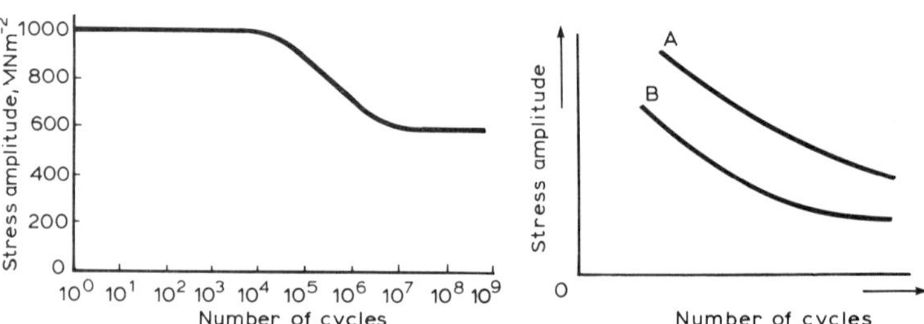

Figure 3.12 *S/N* graph for a nickel-based alloy, Iconel 718

Figure 3.13

4 Creep

Objectives: At the end of this chapter you should be able to:
Describe the essential features of a creep test and the type of results produced.
Explain the terms primary creep, secondary creep, tertiary creep, stress to rupture, isochronous stress/strain graph, creep modulus.
Describe the main factors affecting the creep properties of metals and plastics.
Interpret creep data.

SHORT-TERM AND LONG-TERM BEHAVIOUR

There are many situations where a piece of material is exposed to a stress for a protracted period of time. The stress/strain data obtained from the conventional tensile test refer generally to a situation where the stresses are applied for quite short intervals of time and so the strain results refer only to the immediate values resulting from stresses. Suppose stress were applied to a piece of material and the stress remained acting on the material for a long time–what would be the result? If you tried such an experiment with a strip of lead you would find that the strain would increase with time–the material would increase in length with time even though the stress remained constant. This phenomenon is called *creep,* which can be defined as the continuing deformation of a material with the passage of time when the material is subject to a constant stress.

For metals, other than the very soft metals like lead, creep effects are negligible at ordinary temperatures, but however become significant at higher temperatures. For plastics, creep is often quite significant at ordinary temperatures and even more noticeable at higher temperatures.

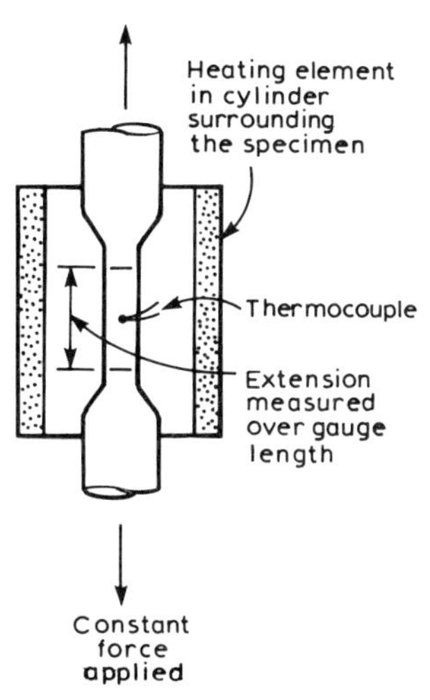

Figure 4.1 A creep test

Figure 4.1 shows the essential features of a creep test. A constant stress is applied to the specimen, sometimes by the simple method of suspending loads from it. Because creep tests with metals are usualy performed at high temperatures a furnace surrounds the specimen, the temperature of the furnace being held constant by a thermostat. The temperature of the specimen is generally measured by a thermocouple attached to it.

Figure 4.2 shows the general form of results from a creep test. The curve generally has three parts. During the *primary creep* period the strain is changing but the rate at which it is changing with time decreases. During the *secondary creep* period the strain increases steadily with time at a constant rate. During the *tertiary creep* period the rate at which the strain is changing increases and eventually causes failure. Thus the initial stress, which did not produce early failure, will result in a failure after some period of time. Such an initial stress is referred to as the *stress to rupture* in some particular time. Thus an acrylic plastic may have a rupture stress of 50 N mm^{-2} at room temperature for failure in one week.

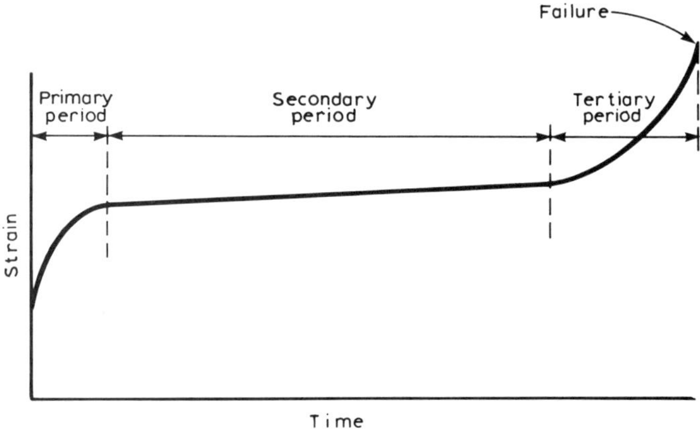

Figure 4.2 Typical creep curve for a metal

FACTORS AFFECTING CREEP BEHAVIOUR WITH METALS

For a particular material, the creep behaviour depends on both the temperature and the initial stress; the higher the temperature the greater the creep, also the higher the stress the greater the creep. *Figure 4.3* shows both these effects. Thus to minimise creep, the conditions need to be low stress and low temperature.

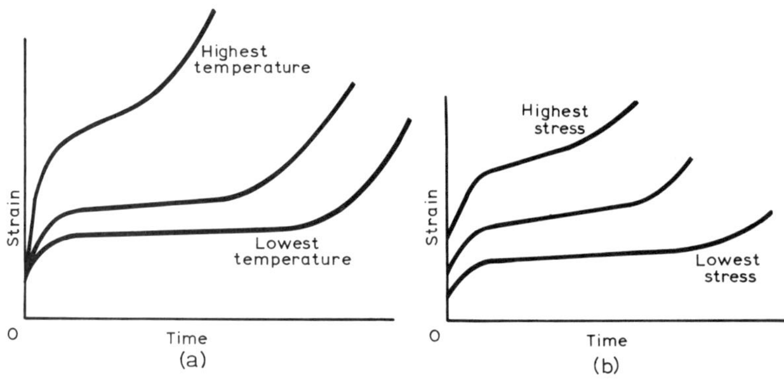

Figure 4.3 Creep behaviour for a material (a) at different temperatures but subject to constant stress (b) at different stresses but subject to constant temperature

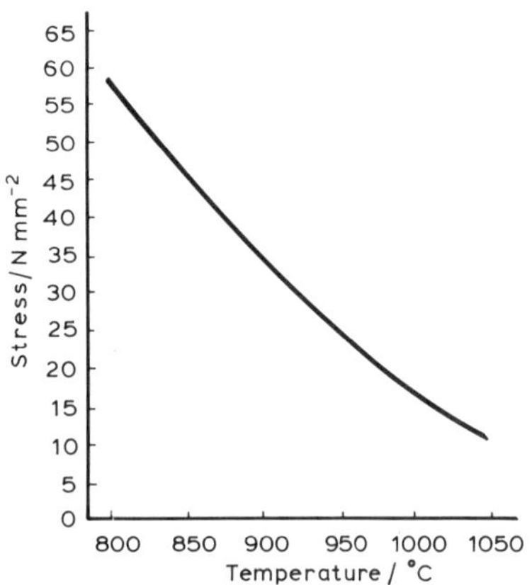

Figure 4.4 Data to give 1% creep in 10 000 h for Pireks 25/20 alloy (0.45% C, 0.8% Mn, 1.2% Si, 20% Cr, 25% Ni). (Courtesy of Darwins Alloy Castings Ltd)

Figure 4.4 shows one way of presenting creep data, indicating the design stress that can be permitted at any temperature if the creep is to be kept within specified limits. In the example given, the limit is 1% creep in 10 000 hours. Thus for the Pireks 25/20 nickel-chrome alloy a stress of 58.6 N mm^{-2} at a temperature of 800°C will produce the 1% creep in 10 000 h. At 1050° a stress of only 10.3 N mm^{-2} will produce the same creep in 10 000 h. *Figure 4.5* shows how stress to rupture the material in 10 000 h varies with temperature. At 800°C a stress of 65.0 N mm^{-2} will result in the Pireks 25/20 alloy failing in 10 000 h; at 1050°C a stress of only 14.5 N mm^{-2} will result in failure in the same time.

Another factor that determines the creep behaviour of a metal is its composition. *Figure 4.6* shows how the stress to rupture different materials in 1000 h varies with temperature. Aluminium alloys fail at quite low stresses when the temperature rises over 200°C. Titanium alloys can be used at higher temperatures before the stress to rupture

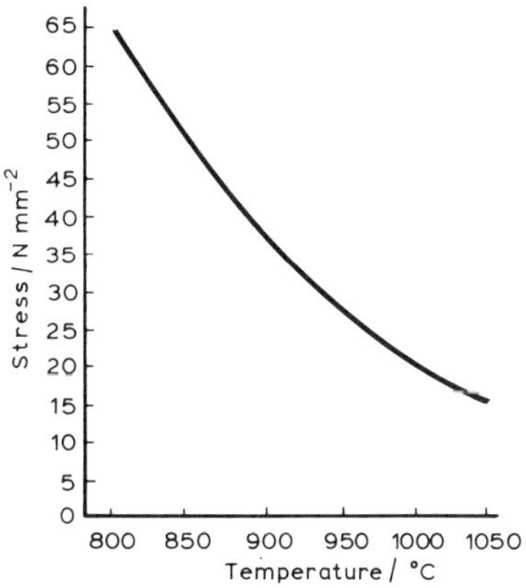

Figure 4.5 Data showing stress to rupture at 10 000 h for Pireks 25/20 alloy. (See also Figure 4.4)

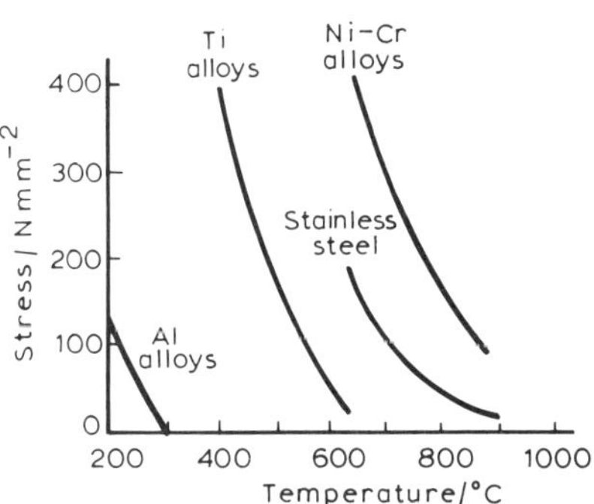

Figure 4.6 Stress to rupture in 1000 h for different materials

drops to very low values, while stainless steel is even better and nickel-chromium alloys offer yet better resistance to creep.

FACTORS AFFECTING CREEP BEHAVIOUR WITH PLASTICS

While creep is significant mainly for metals at high temperatures, creep can be significant with plastics at normal temperatures. The creep behaviour of a plastic depends on temperature and stress, just like metals. It also depends on the type of plastic involved – flexible plastics show more creep than stiff ones.

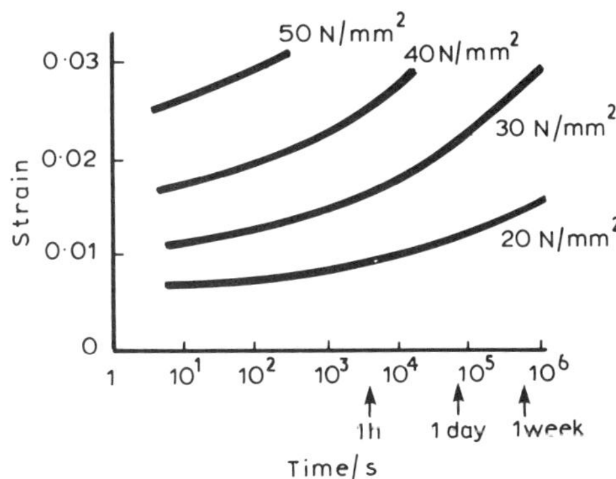

Figure 4.7 Creep behaviour of polyacetal at different stresses. Note logarithmic time scale

Figure 4.7 shows how the strain on a sample of polyacetal at 20°C varies with time for different stresses. The higher the stress the greater the creep. As can be seen from the graph, the plastic creeps quite substantially in a period of just over a week, even at relatively low stresses.

Figure 4.7 is the form of graph obtained by plotting of results derived from creep tests. Based on the *Figure 4.7* graph of strain against time at different stresses a graph of stress against strain for different times can be produced. Thus for a time of 10^2 s, a vertical line drawn on *Figure 4.7* enables the stresses needed for different strains after this time to be read from the graph (*Figure 4.8a*). The resulting stress/strain graph is shown in *Figure 4.8b* and is known as an *isochronous stress/strain graph*. For a specific time the quantity

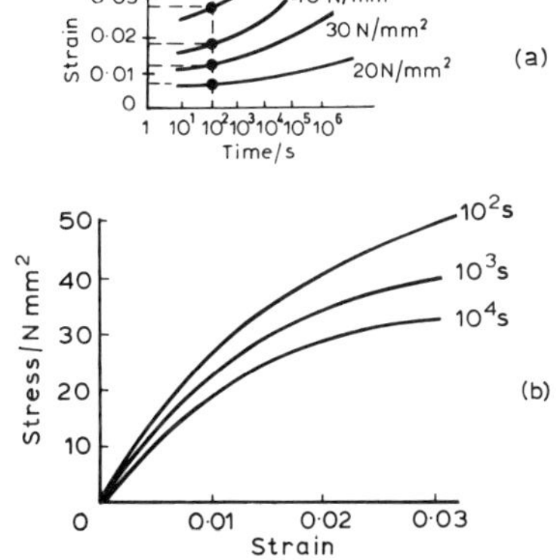

Figure 4.8 (a) Obtaining stress/strain data (b) Isochronous stress/strain graph

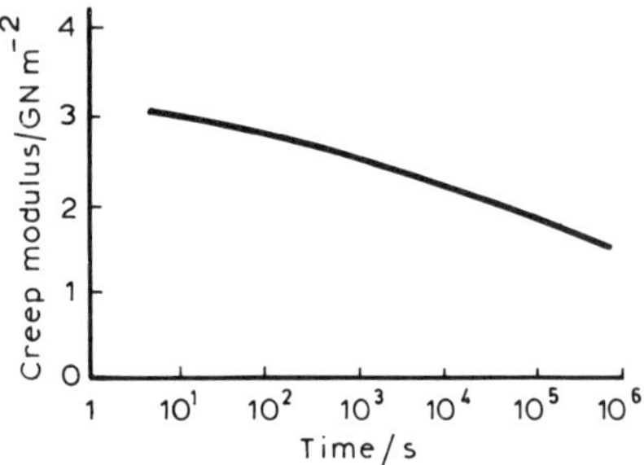

Figure 4.9 Variation of creep modulus with time for polyacetal at 0.5% strain and 20°C

obtained by dividing the stress by the strain for the isochronous stress/strain graph can be calculated and is known as the *creep modulus.* It is not the same as Young's modulus though can it be used to compare the stiffness of plastics. The creep modulus varies both with time and strain and *Figure 4.9* shows how, at 0.5% strain and 20°C it varies with time for the polyacetal described in *Figure 4.7* and *4.8.*

Figure 4.10 shows how the stress to rupture a plastic, Durethan – a polyamide, varies with time at different temperatures. The higher the temperature the lower the stress needed to rupture the material after any particular time.

PROBLEMS

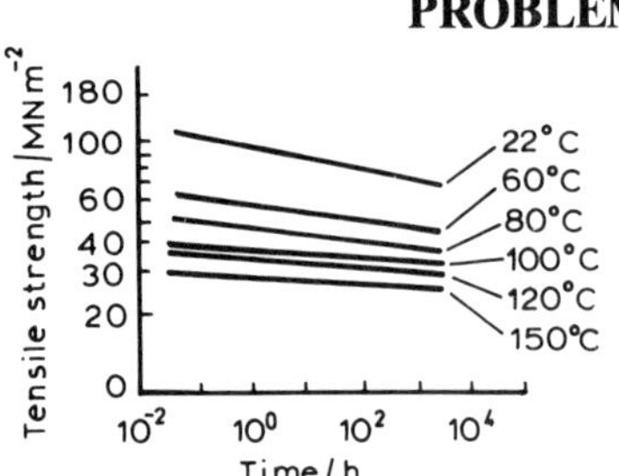

Figure 4.10 Creep rupture graph for Durethan BKV 30 (Courtesy Bayer UK Ltd)

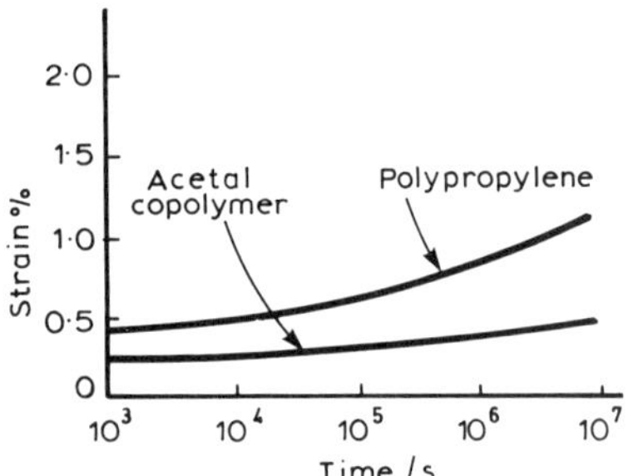

Figure 4.11

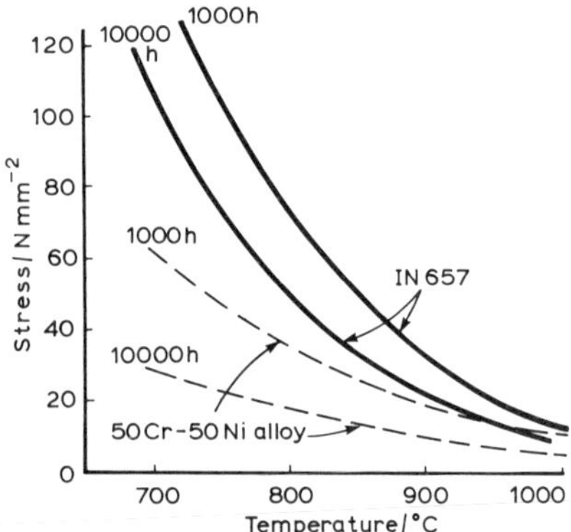

Figure 4.12 Creep rupture data from 'High chromium Cr-Ni alloys to resist residual fuel oil ash corrosion'. (Courtesy of Inco Europe Ltd)

(1) Explain what is meant by 'creep'.

(2) Describe the form of a typical strain/time graph resulting from a creep test and explain the significance of the slope of the graph at its various stages.

(3) Describe the effect of (a) increased stress and (b) increased temperature on the creep behaviour of materials.

(4) For the alloy described in *Figure 4.4* estimate the stress that will result in a 1% creep at a temperature of 900°C after 10 000 h.

(5) For the alloy described in *Figure 4.5,* estimate the stress to rupture at 10 000 h for a temperature of 900°C.

(6) Under what circumstances, would you consider that a metal like that described in *Figures 4.4* and *4.5,* would be necessary? What would you estimate the limiting temperature for use of such an alloy?

(7) Explain how an isochronous stress/strain graph for a polymer can be obtained from creep test results.

(8) Explain the significance of the graph shown in *Figure 4.10* for the creep rupture behaviours of Durethan.

(9) Durethan, as described by *Figure 4.10,* is used for car fan blades, fuse box covers, door handles and plastic seats. How would the behaviour of the material change when the temperature or stress rises?

(10) *Figure 4.11* shows how the strain changes with time for two different polymers when they are subjected to a constant stress. Describe how the materials will creep with time. Which material will creep the most?

(11) *Figure 4.12* shows the stress rupture properties of two alloys, one 50% chromium and 50% nickel, the other (IN 657) 48–52% chromium, 1.4–1.7% niobium, 0.1% carbon, 0.16% nitrogen, 0.2% carbon + nitrogen, 0.5% maximum silicon, 1.0% iron, 0.3% maximum manganese and the remainder nickel. The creep rupture data is presented for two different times, 1000 h and 10 000 h.

(a) What is the significance of the difference between the 1000 h and 10 000 h graphs?

(b) What is the difference in behaviour of the 50 Cr–50 Ni alloy and the IN 657 alloy when temperatures increase?

(c) The IN 657 alloy is said to show 'improved hot strength' when compared with the 50 Cr–50 Ni alloy. Explain this statement.

5 Environmental stability of materials

Objectives: At the end of this chapter you should be able to:
Describe the process of corrosion, explaining such terms as galvanic corrosion, galvanic series, demetallification, graphitisation, concentration cell, stress corrosion, differential aeration cells.
Describe methods of corrosion prevention involving selection of materials, selection of design, modification of the environment and protective coatings.
Describe the factors affecting the stability of polymers and ceramics.
Identify potential corrosion situations.

CORROSION

The car owner can rightly be concerned about rust patches appearing on the bodywork of the car, as the rust not only makes the bodywork look shoddy but indicates a mechanical weakening of the material. The steel used for the car bodywork has thus changed with time due to an interaction between it and the environment. The possibility of such corrosion is therefore a factor that has to be taken into account when a material is selected for a particular purpose.

Most metals react, at moderate temperatures, only slowly with the oxygen in the air. The result is the build up of a layer of oxide on the surface of the metal, which can insulate the metal from further reaction with the oxygen. Aluminium is an example of a metal that builds up a protective oxide layer which is a very effective barrier against further oxidation. If oxidation were the only reaction which tended to occur between a metal and its environment then probably corrosion would present few problems. The presence of moisture in the environment can very markedly affect corrosion, as can the presence of chemically active pollutants.

Iron rusts when the environment contains both oxygen and moisture, but not with oxygen alone or moisture alone. Iron nails kept in a container with dry air do not rust, nor if they are kept in oxygen-free, e.g. boiled, water. But in moist air they rust readily.

Chemically-active pollutants in the environment can have a marked effect on corrosion, especially those that are soluble in water. Man-made pollutants such as the oxides of carbon and sulphur, produced in the combustion of fuels, dissolve in water to give acids, which readily attack metals and many other materials. Marine environments also are particularly corrosive, due to the high concentrations of salt from the sea. The sodium chloride reacts with metals to produce chlorides of the metals which are soluble in water and thus cannot act as a protective layer on the surface of a metal as a non-soluble oxide may do. The salt may also destroy the protective oxide layer that has been acquired by a metal.

GALVANIC CORROSION

A simple electrical cell could be just a plate of copper and one of zinc (*Figure 5.1*) dipping into an *electrolyte,* a solution that conducts electricity. Such a cell gives a potential difference between the two

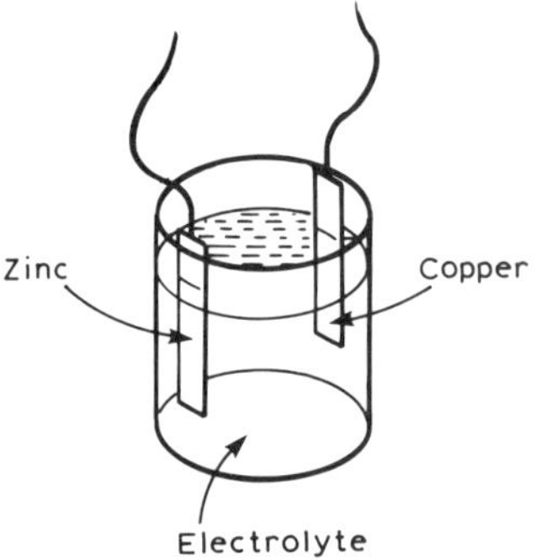

Figure 5.1 A simple cell

metals – it can be measured with a voltmeter or used to light a lamp. Different pairs of metals give different potential differences. Thus a zinc-copper cell gives a potential difference of about 1.1 V, and an iron-copper cell about 0.8 V. A zinc-iron cell gives a potential difference of about 0.3 V; this value is however the difference between the potential differences of the zinc-copper and iron-copper cells. It is as though we had a cell made up with zinc-copper-iron.

By tabulating values of the potential differences between the various metals and standard, a table can be produced from which the potential differences between any pair of metals can be forecast. The standard used is hydrogen and the following are the potential differences relative to hydrogen for a number of metals.

Metal	*Potential difference*/V
Gold	+1.7
Silver	+0.80
Copper	+0.34
Hydrogen	0.00
Lead	−0.13
Tin	−0.14
Nickel	−0.25
Iron	−0.44
Zinc	−0.76
Aluminium	−1.67
Magnesium	−2.34
Sodium	−2.71

The table gives, for a silver-aluminium cell, a potential difference of 2.47 V, that is +0.80 – (−1.67) V. The metal highest up the table, as shown above, behaves as the negative electrode of the cell while the metal lowest in the table is the positive electrode. The term *cathode* is used for the negative electrode and *anode* for the positive electrode. Thus for the silver-aluminium cell the silver is the negative electrode and the aluminium the positive electrode.

With a copper-zinc cell it is found that the copper electrode remains unchanged after a period of cell use but the zinc electrode is badly corroded. With any cell the anode is corroded and the cathode not affected. The greater the cell potential difference the greater the corrosion of the anode.

The above table lists what is called the *electromotive series* or *galvanic series.* Tables of such series are available for metals in various environments. Such tables are of use in assessing the possibilities of corrosion when two different metals are in electrical contact, either directly or through a common electrolyte.

If a copper pipe is connected to a galvanised steel tank, perhaps the cold water storage tank in your home, a cell is created (*Figure 5.2*) and corrosion follows. Galvanised steel is zinc-coated steel; there is thus a copper-zinc cell. With such a cell the copper is the cathode and the zinc the anode; the zinc thus corrodes and so exposes the iron. Iron-copper is also a good cell with the iron as the anode. The result is corrosion of the iron and hence the overall result of connecting the copper pipe to the tank is likely to be a leaking tank.

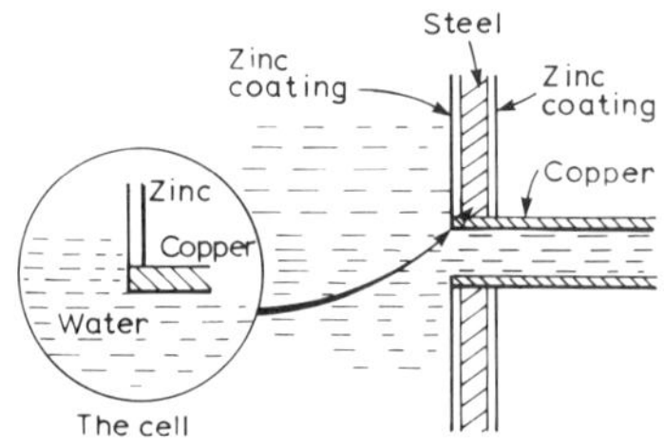

Figure 5.2 Copper-zinc cell for copper pipe connection to galvanised steel water tank

Stainless steel and mild steel form a cell, the mild steel being the anode. Thus if a stainless steel trim, on say a car, is in electrical contact with mild steel bodywork, then the bodywork will corrode more rapidly than if no stainless steel trim were used. The electrical connection between the stainless steel and the mild steel may be through water gathering at the junction between the two. With oxygen-free clean water the cell potential difference may be only 0.15 V, but if the water were sea water containing oxygen the potential difference could become as high as 0.75 V. So the use of salt in Britain, to melt ice on the roads, leads to greater corrosion of cars.

It is not only with two separate metals that cells can be produced and corrosion occur, as galvanic cells can be produced in a number of ways. An alloy or a metal containing impurities can give rise to galvanic cells within itself. For example, brass is an alloy of copper and zinc, a copper-zinc cell has the zinc as the anode and thus the zinc corrodes. The effect is called *dezincification,* one example of *demetallification.* After such corrosion the remaining metal is likely to be porous and lacking in mechanical strength (see Table 5.1).

A similar type of corrosion takes place for carbon steels in the pearlitic condition, the cementite in the steel acting as the cathode and the ferrite as the anode. The ferrite is thus corroded.

Cast-iron is a mixture of iron and graphite, the graphite acting as the cathode and the iron as the anode. The iron is thus corroded, the effect being known as *graphitisation.*

Table 5.1 Typical manufacturer's information concerning dezincification (Courtesy of McKechnie Metals Ltd)

Dezincification

Brasses containing 80% copper or higher are rarely corroded in this manner, and almost complete immunity is obtained by the addition of 0.02–0.06% arsenic in the alpha brasses of lower copper contents.

Until recently, no satisfactory answer was available for the alpha-beta duplex brasses. Research has established an alloy of this type, Alloy CZ132, now covered by BS 2872 and BS 2874, which is dezincification resistant.

Care is necessary when using other duplex alpha-beta brasses in hot water applications where the water has characteristics prone to cause dezincification, and these alloys are not recommended for use in sea water.

Variations in concentration of the electrolyte in contact with a metal can lead to corrosion. That part of the metal in contact with the more concentrated electrolyte acts as a cathode while the part in contact with the more dilute electrolyte acts as an anode and so corrodes most. Such a cell is known as a *concentration cell.*

Another type of concentration cell is produced if there are variations in the amount of oxygen dissolved in the water in contact with a metal. That part of the metal in contact with the water having the greatest concentration of oxygen acts as a cathode while the part of the metal in contact with the water having the least concentration acts as an anode and so corrodes most. A drop of water on a steel surface is likely to have a higher concentration of dissolved oxygen near its surface where it is in contact with air than in the centre of the drop (*Figure 5.3*). The metal at the centre of the drop acts as an anode and so corrodes most.

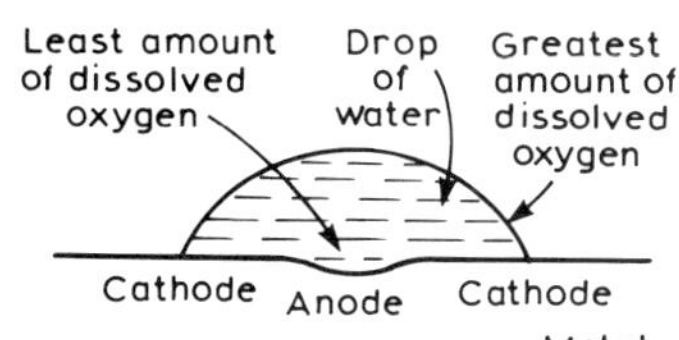

Figure 5.3 Example of a concentration cell

Variations in stress within a metal or a component can lead to the production of cells and hence corrosion. A component which has part of it heavily cold-worked and part less-worked will contain internal stresses which can result in the heavily-worked part acting as an anode and the less-worked part as a cathode. Therefore the heavily cold-worked part corrodes most.

Table 5.2 Typical manufacturer's information concerning stress corrosion with brass (Courtesy of McKechnie Metals Ltd).

'Season cracking' and stress corrosion cracking

Both these forms of cracking are due to the combined effect of corrosion and stress. In 'season cracking' (a term used specifically for stress corrosion in brass), residual stress in the metal, combined with extremely minute traces of ammonia or ammonia derivatives in moist air, can result in cold-worked brass cracking.

Failure is time-dependent and the higher the residual stress or the higher the concentration of ammonia, the shorter the time necessary for the development of cracks. Brasses containing less than 15 per cent zinc are rarely prone to season cracking.

Prevention of failure by season cracking can be accomplished by either (a) low-temperature annealing the brass at 300°C, which relieves the stresses without altering significantly the level of tensile strength, or (b) redistributing the stresses by a mechanical method such as reeling.

Stress corrosion cracking can occur with residual or applied stress in conjunction with corrosion. This form of failure is neither confined to one kind of material nor to one form of corrosive medium, but is encountered with steels, aluminium alloys, magnesium alloys, copper base alloys, plastics and rubber. The selection of materials for service where they are likely to be exposed to corrosive environments is thus often critical.

CORROSION PREVENTION

Methods of preventing corrosion, or reducing it, can be summarised as:

(1) Selection of appropriate materials.
(2) Selection of appropriate design.
(3) Modification of the environment.
(4) Use of protective coatings.

In selecting materials, care should be taken not to use two different materials in close proximity, particularly if they are far apart in the galvanic series, giving a high potential difference between them. The material which acts as the anode will be corroded in the appropriate environment. It is not desirable to connect copper pipes to steel water tanks. Steel pipes to a steel tank would be better.

However there are situations where the introduction of a dissimilar metal can be used for protection of another metal. Pieces of magnesium or zinc placed close to buried iron pipes can protect the pipes in that the magnesium or zinc acts as the anode relative to the iron which becomes the cathode. The result is corrosion of the magnesium or zinc and not the pipe. Such a method is known as *galvanic protection.*

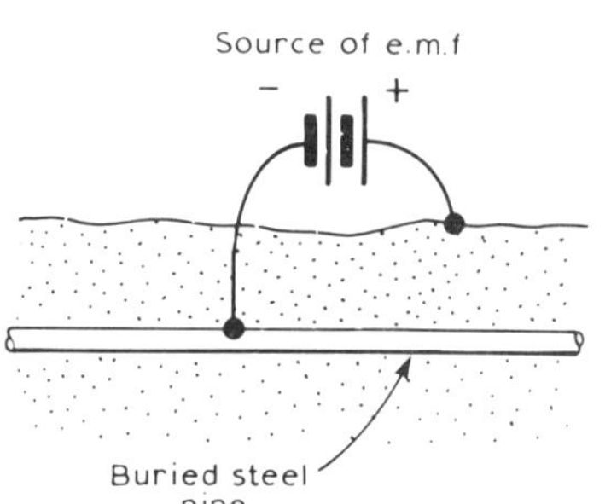

Figure 5.4 Corrosion prevention using an applied e.m.f.

The steel hull of a ship can be protected below the water line by fixing pieces of magnesium or zinc to it. The steel then behaves as the cathode, the magnesium or zinc becoming the anode and so corroding. Another way of making a piece of metal act as a cathode and so not corrode, is to connect it to a source of e.m.f. in such a way that the externally applied potential difference makes the metal the cathode in an electrical circuit (*Figure 5.4*).

Selection of an appropriate metal for a specific environment can do much to reduce corrosion. These are the properties of some metals in different environments.

(1) *Copper*
When exposed to the atmosphere copper develops an adherent protective layer which then insulates it from further corrosion. Copper is used for water pipes, e.g. domestic water supply pipes, as it offers a high resistance to corrosion in such situations.

(2) *Copper alloys*
Demetallification can occur. Some of the alloying metals, however, can improve the corrosion resistance of the copper by improving the development of the adherent protective surface layer.

(3) *Iron and steel*
This is corroded readily in many environments and exposure to sea water can result in graphitisation. Stress corrosion can occur in certain environments.

Table 5.3 Comparison of zinc coatings (Courtesy of the Zinc Development Association)

	Sherardizing	*Hot Dip Galvanizing*	*Electro-Plating*	*Zinc Spraying*	*Zinc Rich Paints*
Adhesion	Good. The zinc coating is alloyed with the basis metal	Good. The zinc coating is alloyed with the basis steel	Good, comparable with other electro-plated coatings	Good mechanical interlocking provided the abrasive grit blasting pretreatment is done correctly	Good. Abrasive grit blasting of the steel gives best results
Continuity and Uniformity	Continuous and very uniform even on threaded and irregular shaped parts	Generally uniform. Some excess zinc at drainage points	Uniform within limitations of 'throwing power' of the plating bath	Depends on operators skill	Depends on operator's skill
Thickness	Thickness variable at will, generally 15 to 40 µm (0.0006 in.–0.0016 in.). Thicker coatings also possible	Generally 45 to 125 µm (0.0018 in.–0.005 in.) on products. 25 µm (0.001 in.) on sheet. Thicker coatings also possible	Thickness variable at will, generally 2–25 µm (0.0001 in.–0.0010 in.). Thicker coatings are possible but generally uneconomic	Thickness variable at will dependant on operator's skill, generally 100 to 200 µm (0.004 in.–0.008 in.). Thicker coatings can be applied	Up to 40 µm (0.0016 in.) of paint (and more with special formulations) can be applied with one coat
Formability and Mechanical Properties	Usually applied to finished articles. The thinner coatings can be formed but heavy coatings can be brittle on bending. Excellent abrasion resistance, the hardest of all zinc coatings	Conventional coatings applied to finished articles, not formable: alloy layer is abrasion resistant but brittle on bending. Galvanized sheet has l ttle or no alloy layer and is readily formed	Can be formed if required but small parts usually finished before plating	Usually applied to finished articles, forming not required. Not highly abrasion resistant	Painted sheet can be formed. Poor abrasion resistance
After Treatments	Good key for paint. Phosphate and Chromate conversion coatings used to prevent wet storage stain	Etching treatment or weathering required before painting. Chromate conversion coatings used to prevent wet storage stain	Treatment required before painting. Chromate conversion coatings used to prevent wet storage stain	Good key for paint	Can be used alone, or as primer under conventional paints
Other Considerations	Limited by size of containers. Generally used for fairly small complex components. Useful when close control of tolerances important. Unsuitable for large bulky items, tanks, etc. Narrow parts up to 20 ft. long can be treated.	Limited by size of bath available. Parts up to 90 ft. long and fabrications 55 ft. x 6 ft. x 15 ft. can be dipped at specialized works. Care required at design stage for best results. Continuous galvanized wire and strip (up to 4 ft. wide) available.	Limited by size of bath available. Generally used for simple fairly small components suitable for barrel plating of for continuous sheet and wire	No size limitations but access difficulties may limit applications Not generally suitable for small parts.	No size limitations. Suitable for anything that can be painted by brush, spray or dip.
British Standards applicable	B.S. 4921 : 1973	General Work: B.S. 729 : 1971	General Work: B.S. 1706 : 1960 Threaded Components: B.S. 3382 : 1961	B.S. 2569 : 1964	None at present

Table 5.4 Typical manufacturer's information concerning surface treatments for aluminium. (Courtesy of Alcan Extrusions Ltd)

Description of finish	Characteristics and uses
Mill finish	
The untreated surface of the material as manufactured.	The natural oxide film gives adequate protection for applications where lowest cost and no maintenance are required, and where a change in surface appearance is acceptable. For example; patent glazing bars on roofs, many greenhouse and transport sections.
Painted	
Electrophoretically applied organic coating followed by stoving.	The electrophoretic process gives complete coverage with excellent uniformity of thickness. Principally used in windows and doors.
Anodised—Natural	
For external applications anodising should be carried out to the recommendations of BS3987:1966; for internal applications BS1615:1972 should be referred to*. With appropriate pretreatments, satin and other surface effects can be obtained.	Anodising produces a relatively thick oxide film that preserves the smooth surface of the metal; used when appearance is important, or in engineering applications when a harder finish is required.
Anolok	
An Alcan process, giving a more durable light-fast finish than the conventional dyed colour anodising. The chemically-stable metallic oxides that provide the colour are deposited at the base of the anodic film.	The light-fast colours – various bronzes and black – and the excellent durability of the finish make Anolok suitable for decoration where good appearance is essential. The range is extended by the use of polished or other finishes.
Polished	
Sections are linished or buffed mechanically using a polishing compound. Polished sections are anodised after polishing to preserve the finish.	Polishing is particularly recommended for architectural applications, such as shop fronts and entrance units for which the highest standard of finish is required.

*The following grades of coating thickness are supplied according to BS1615 (the suffix number represents minimum average film thickness in microns): AA25, AA20, AA15, AA10, AA5. They are specified for architectural and industrial use as follows:
Grade AA25 External applications cleaned at least four times a year.
Grades AA20, AA15 External applications cleaned more than four times a year.

(4) *Alloy steels*
The addition of chromium to steel can improve considerably the corrosion resistance by modifying the surface protective film produced under oxidising conditions. The addition of nickel to an iron-chromium alloy can further improve corrosion resistance and such a material can be used in sea water with little corrosion resulting. Austenitic steels, however, are susceptible to stress corrosion in certain environments. Iron-nickel alloys have good corrosion resistance and such alloys with 20% nickel and 2 to 3% carbon are particularly good for marine environments.

(5) *Zinc*
Zinc can develop a durable oxide layer in the atmosphere and then becomes resistant to corrosion.

Table 5.4 (cont.)

Maintenance	Life of finish
Surface roughening of mill-finish aluminium with weathering cannot be avoided. The metal will retain its original colour with very little attention in rural areas, but will darken more quickly in urban surroundings due to heavier atmospheric pollution. It is difficult to prevent this, even with regular washing, and equally difficult to restore the surface to its original condition when it has been heavily discoloured.	Surface darkens with weathering, but this does not affect the strength and performance of the metal.
Periodic washing with water to which may be added a mild detergent. Abrasives must not be used.	Life of stoved paint system. No rusting or lifting at scratches or chips.
Periodic washing with water to which may be added a mild detergent. Abrasives must not be used.	Assuming reasonable environment and frequent cleaning, the anodic film should remain substantially intact for a period of 25 years. In some cases, localised deterioration of the film may occur earlier.
Periodic washing with water to which may be added a mild detergent. Abrasives must not be used.	The life of the film is as for 'Natural Anodising' above. The colour is durable and completely light-fast and will not fade during the lifetime of the anodic film.
Periodic washing with water to which may be added a mild detergent. Abrasives must not be used.	The life of the film is as for 'Natural Anodising' above.

Grades AA15, AA10, AA5 For interior applications, the thicker coatings being specified for greater wear (e.g. handrails) and the thinner for products susceptible to finger marking (e.g. partitions). Externally, thin coatings are used for special applications such as bright trim on vehicles.

(6) *Aluminium*

Readily developing a durable protective surface film, aluminium is then resistant to corrosion. Aluminium alloys are subject to stress corrosion.

The avoidance of potential crevices (*Figure 5.5*) which can hold water or some other electrolyte and so permit a cell to function, perhaps by bringing two dissimilar metals into electrical contact or by producing a concentration cell, should be avoided in design. Suitable design can also do much to reduce the incidence of stress corrosion.

Corrosion can be prevented or reduced by modification of the environment in which a metal is situated. Thus in the case of a packaged item an impervious packaging can be used so that water vapour cannot come into contact with the metal. Residual water

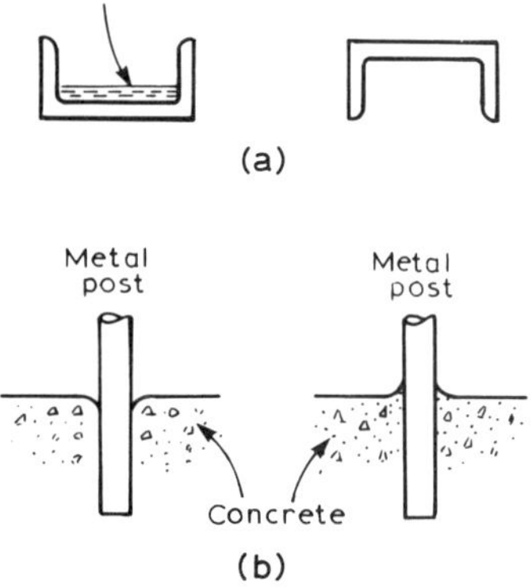

Figure 5.5 Ways of reducing corrosion by preventing water from collecting, (a) Simple inversion (b) Using a fillet to eliminate crevice

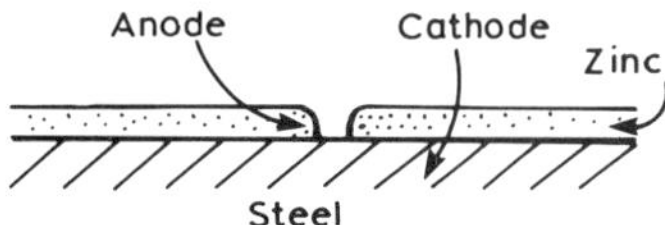

Figure 5.6 Galvanised steel

vapour within the package can be removed by including a dessicant such as silica gel, within the package.

Where the environment adjacent to the metal is a liquid it is possible to add certain compounds to the liquid so that corrosion is inhibited, such additives being known as *inhibitors.* In the case of water-in-steel radiators or boilers, compounds which provide chromate or phosphate ions may be used as inhibitors, as they help to maintain the protective surface films on the steel.

One way of isolating a metal surface from the environment is to cover the metal with a protective coating, which can be impervious to oxygen, water or other pollutants. Coatings of grease, with perhaps the inclusion of specific corrosion inhibitors, can be used to give a temporary protective coating. Plasticised bitumens or resins can be used to give harder but still temporary coatings, while organic polymers or rubber latex can be applied to give coatings which can be stripped off when required.

One of the most common coatings applied to surfaces in order to prevent corrosion is paint, different types having different resistances to corrosion environments. Thus some paints have a good resistance to acids while others are good for water.

Steel components can be protected by dipping them in molten zinc to form a thin surface coating of zinc on the steel surface. This method is known as *galvanising.* The zinc acts as an anode with the steel being the cathode (*Figure 5.6*), and the zinc corrodes, rather than the steel, when the surface layer is broken. Small components can be coated with zinc by heating them in a closed container with zinc dust. This process is called *sherardising.*

Other metals can be used to coat steel, often by means of electroplating; thus nickel-plated steel offers some protection from the environment. Often a layer of chromium is applied to a steel surface over a base coat of nickel; the chromium layer is quite resistant to corrosion. Such coatings though offering protection are not so effective as zinc when the layer is broken, the zinc being sacrificed in place of the steel.

Steel surfaces are often treated with phosphoric acid or solutions containing phosphate ions, resulting in the formation of a phosphate coating, the process being known as *phosphating.* The coating bonds well with the surface and though giving some corrosion protection is generally used as a precursor for other coatings, perhaps paint.

Several metals can have corrosion-resistant surface layers produced by the application of solutions of chromates to their surfaces. A steel surface which has had the phosphating treatment may then be subject to a chromating treatment, the result being a good corrosion resistant surface layer on the metal.

Aluminium in the atmosphere has generally an oxide surface layer which offers some corrosion resistance and can be thickened by an electrolytic process. The treatment is known as *anodising.*

THE STABILITY OF POLYMERS

When compared with metals, polymers are relatively stable. They are largely unaffected by weak acids or alkalis, oils and greases, but may be dissolved by some organic solvents. Exposure to the environment, i.e. the weather, and ultraviolet radiation, can result in some deterioration. This may show itself as a loss of colour of the surface or its finish and there can also be a loss in strength.

Thermoplastics deteriorate with an increase in temperature. Polyethylene, for instance, is limited to use below about 60°C. Thermosetting materials are relatively unaffected by reasonable increases in temperature.

Table 5.5 Typical manufacturer's information concerning the environmental stability of a plastic (Courtesy of Bayer UK Ltd)

Properties of Novodur

ABS polymers are available in numerous varieties with properties tailored to specific requirements. Novodur offers special advantages with regard to the ratio between hardness, rigidity and impact strength, having high heat deflection temperature and chemical resistance as well as outstanding surface quality.

Resistance to chemicals

Novodur is highly resistant to chemicals. It is not affected by alkalis, dilute organic and inorganic acids, aliphatic hydrocarbons (e.g. white spirit), nor by most oils and fats. However, aromatics, ketones, ethers, esters and chlorinated hydrocarbons cause swelling or superficial dissolution, so these solutions must not come into contact with Novodur.

Resistance to weathering

The elastomeric phase of the ABS polymers, which consists of butadiene rubber, is sensitive to oxygen and ultraviolet radiation. Yellowing and other ageing effects are the consequence.

Novodur is stabilised against ageing by atmospheric oxygen. This stabilisation is sufficient to ensure that the mouldings remain serviceable for many years when used indoors.

Under the influence of sunlight, however, articles in white or pastel shades show slight yet perceptible yellowing. The use of ABS polymers outdoors is limited by the ageing process.

Table 5.6 Typical manufacturer's information concerning the environmental stability of a rubber (Courtesy of Du Pont (UK) Ltd).

Neoprene synthetic rubber

Resistance to deterioration from waxes, fats, oils, greases and many other petroleum products is one of the best-known properties of Neoprene. In fact, Neoprene was developed originally as an oil-resistant substitute for natural rubber and is still widely used for this purpose. Its use is limited, however, to non-aromatic hydrocarbons and it will not withstand chlorinated solvents.

Generally speaking, Neoprene products show very little change in properties or appearance when exposed to alkalies, dilute mineral acids, or inorganic salt solutions. However, acid and salt solutions of a highly oxidising nature will cause surface deterioration and loss of strength. Unlike natural rubber and other general-purpose elastomers, Neoprene gives excellent service in contact with aliphatic hydrocarbons, aliphatic hydroxy compounds and most Freon refrigerants. In contact with these fluids, Neoprene displays minimum swelling, relatively little loss of strength and virtually complete recovery of initial properties after removal of the liquid by drainage or evaporation.

Properly compounded black Neoprene has high resistance to ozone, sun and weather.

THE STABILITY OF CERAMICS

Ceramics are relatively stable when exposed to the atmosphere, though the presence of sulphur dioxide in the atmosphere and its subsequent change to sulphuric acid can result in deterioration of ceramics. Thus building materials such as stone and brick can be severely damaged by exposure to industrial atmospheres in which sulphur dioxide is present.

Damage to ceramics may also result from the freezing of water which has become absorbed into pores of the material and from thermal shock. The low thermal conductivity of ceramics can result in large thermal gradients being set up and hence considerable stresses. This can lead to flaking of the surface. Ceramics used as furnace linings may well be affected by molten metals or slags. Furnace linings need to be chosen with care.

PROBLEMS

(1) It has been observed that cars in a dry desert part of a country remain remarkably free of rust when compared with cars in a damp climate such as England. Offer an explanation for this.

(2) Why does the de-aeration of water in a boiler reduce corrosion?

(3) What criteria should be used if corrosion is to be kept to a low value when two dissimilar metals are joined together?

(4) Aluminium pipes are to be used to carry water into a water tank. Possible materials for the tank are copper or galvanised steel. Which material would you advocate if corrosion is to be minimised?

(5) It is found that for a junction between mild steel and copper in a sea water environment that the mild steel corrodes rather that the copper. With a mild steel — aluminium junction in the same environment it is found that the aluminium corrodes more than the mild steel. Explain the above observations.

(6) Pieces of magnesium placed close to buried iron pipes are used to reduce the corrosion of the iron. Explain.

(7) Corrosion of a stainless steel flange is found to occur when a lead gasket is used. Explain.

(8) Propose a method to give galvanic protection for a steel water storage tank.

(9) Why should copper piping not be used to supply water to a galvanised steel water storage tank?

(10) What are the main ways galvanic cells can be set up in metals?

(11) Compare the use of zinc and tin as protective coatings for steel.

(12) Compare the relative merits of the corrosion prevention processes of Sheradising, hot dip galvanising, electro-plating with zinc, zinc spraying and coating with zinc rich paints. Table 5.3, reprinted by courtesy of Zinc Alloy Rust-proofing Co. Ltd should be of help.

(13) The following note appeared in the Products and Techniques section of the December 1977 issue of *Engineering*. Explain the purpose and mode of action of the anodes referred to in the note.

"Corrosion anodes
Solid anodes for the cathodic protection of water tanks are now available in a variety of sizes and three types — freestanding, suspended and weld attached. No prior preparation is required before installation, and protection begins at once. Being non-toxic, the Metalife anodes can be used in tanks for drinking water. Belzona Molecular Metalife Ltd, Harrogate."

(14) What further information do you think you should need to assess the merits of the anodes referred to in the previous question? Do you think it would matter what type of material the water tank was made of?

(15) In an article on underwater equipment design the author states that good design required the avoidance of crevices and high stress concentrations and that also the number of different metals should be kept to a minimum. Give explanations for these criteria.

Engineering, April 1976

10 ways to help corrosion

Corrosion is an excellent way of gradually ruining equipment. Carefully calculated corrosion assistance can also produce spectacular failures. Usually corrosion is only considered seriously after it has happened. This may take some time, so here are some hints for the designer who just can't wait. There are many ways to help corrosion, but K Farrow, of the Fulmer Research Institute, recommends these top ten steps to disaster.

1 Raise the temperature. This increases the rate of corrosion reactions. Suitable design will give temperature differences producing thermal stresses and hot spots. This should produce localised corrosion.

Protective layers of corrosion product may develop and reduce high temperature corrosion rates. These layers are, however, often brittle and generally have a different coefficient of thermal expansion to the base metal. Rapid temperature fluctuations can therefore cause spalling of the protective layers ensuring continuous corrosion.

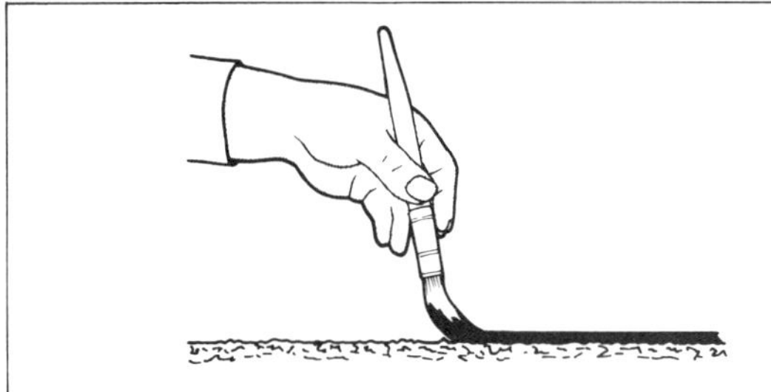

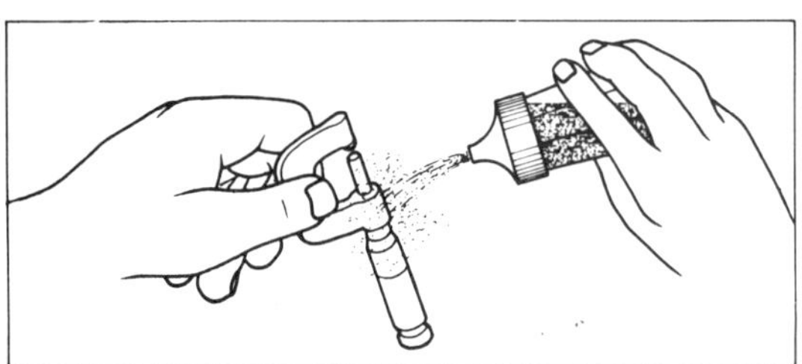

2 Add a supply of chloride. Common salt is a cheap and readily available source. This increases the electrical conductivity of the water and speeds up corrosion processes.

Chloride can also cause breakdown of the protective oxide film on stainless steel and corrosion resistant aluminium alloys. This gives localised corrosion and pitting. The pits reduce the section thickness and also act as stress raisers. As a further bonus the pits may act as initiators for stress corrosion cracking (see 6) or corrosion fatigue (see 4).

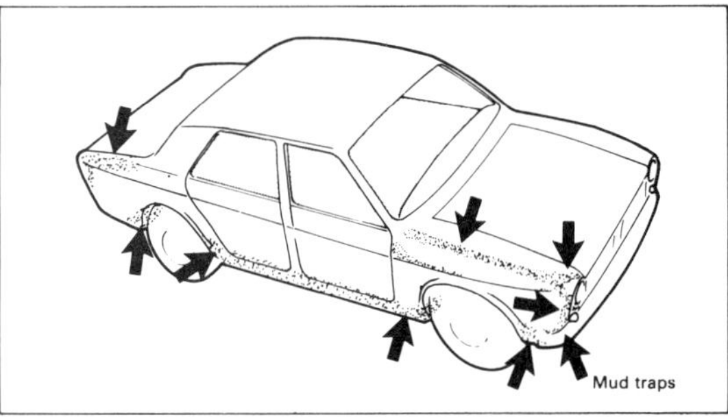

3 Add water. This gives better corrosion than relying on atmospheric moisture.

A spray system is probably best but periodic rinsing or immersion are good substitutes. Where none of these is possible, moisture traps can usually be included in the design. These include cracks, crevices and undrainable regions. Deliquescent materials can also be used to attract moisture.

If all else is impractical then provide wood shavings, or soil, to ensure that any water splashed on to the structure is retained.

4 Metal parts can generally be persuaded to fail prematurely if subjected to fatigue loading. Oscillating levels, vibrations or balances can all be used.

More important, however, the fatigue properties of materials are generally poorer in more corrosive environments (corrosion fatigue). This will obviously be apparent where pitting corrosion (see 2) has produced stress concentrations. Any more agressively corrosive environment should, however, speed up time to failure.

5 Design in a few rubbing surfaces and include a supply of abrasive particles.

General corrosion tends to slow down or halt as a layer of corrosion product forms. If this is constantly eroded or abraded away then corrosion can proceed unhindered.

This can easily be done by specifying a few under-torqued fasteners and a dusty atmosphere. Save money by dispensing with air cleaners and other filters.

With pipework, sharp changes in direction and section can reduce service lifetime. In time collapsing bubbles can hammer their way through steel piping.

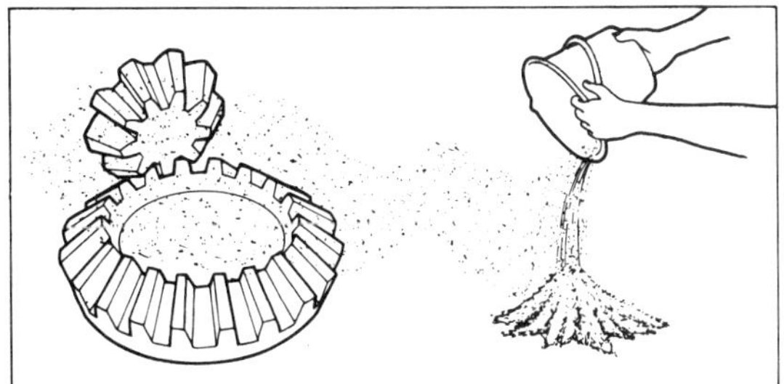

6 Don't consider stress corrosion cracking. This form of corrosion is insidious, neat and potentially catastrophic. Essentially required are a suitably stressed component and a specific environment. The component can then break 'unexpectedly' and cause untold havoc.

This is, however, a sophisticated form of corrosion and requires more technical expertise than most. Consult an expert.

7 Ignore the more esoteric forms of corrosion—on which a corrosion consultant could advise.

Hydrogen can cause dramatic embrittlement and fracture particularly in very high strength steels. The gas itself or acids can be used.

Liquid metals can cause either accelerated corrosion or embrittlement and fracture. As an example mercury can have a ruinous effect upon aluminium alloys. The result can be either embrittlement (particularly with high strength aluminium alloys) or corrosion so rapid that the corrosion product can be seen to grow.

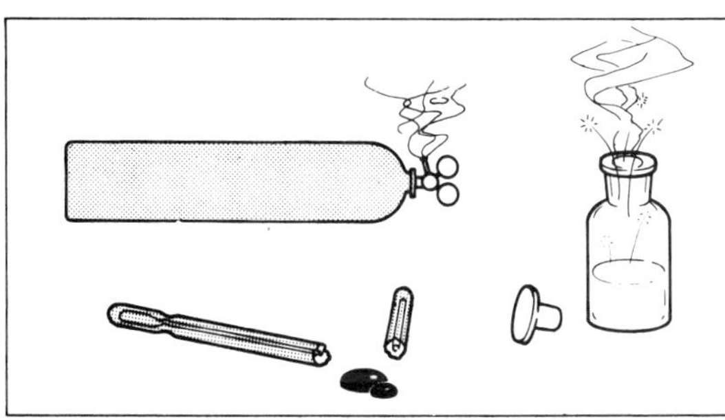

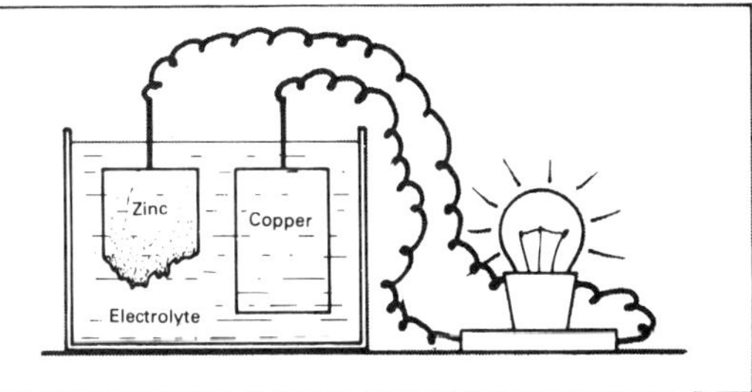

8 Use lots of different metals and alloys joined together. Preferably select metals and alloys far apart in the list below. (A galvanic series). The lower one should then corrode more rapidly.

If the rapidly corroding metal (lower in the list) is smaller in area, so much the better. It will then corrode even faster. As an example use mild steel rivets to join stainless steel sheet. The rivets should then corrode away extremely rapidly giving easy disassembly.

Gold	Noble or resistant to corrosion.
Silver	
Stainless steel (passive) with protective oxide layer.	
Bronzes	
Copper	
Brasses	
Tin	
Lead	
Stainless steel (active) with non protective layer (see 2)	
Cast iron	
Steel	
Aluminium	
Zinc	
Magnesium	Base or readily corroded

9 Provide electrical currents to give accelerated corrosion. The anodic parts of a structure can then be designed to crumble away. Hydrogen produced in cathodic regions could also be used to produce embrittlement (see 7). If a suitable source of impressed current cannot be included in the design then site the equipment adjacent to another electric power source.

10 Protective coatings offer at least two ways to help corrosion.

If cost is important do not specify any paint or protective coatings.

If money is no object specify an expensive coating but don't waste time on preparation, application or maintenance. As examples, steel can always be allowed to rust heavily and then paint be applied onto the rust. Similarly plating can generally be ruined by application to greasy metal.

6 Forming processes with metals

Objectives: At the end of this chapter you should be able to:
Distinguish the four main categories of forming, i.e. casting, manipulative processes, powder techniques and cutting/grinding, and the types of products that can be produced by each.
Identify the criteria which need to be considered before a choice of forming process can be made.

THE MAIN PROCESSES

The range of forming processes can be divided broadly into four categories:

(1) Casting, shaping a material by pouring the liquid material into a mould.
(2) Manipulative processes, shaping a material by plastic deformation processes.
(3) Powder techniques, producing a shape by compacting a powder.
(4) Cutting and grinding, producing a shape by metal removal.

The choice of forming process will depend on a number of factors:

(1) The quantity of items required.
(2) The dimensional accuracy required.
(3) The surface finish required.
(4) The size of the items, both overall size and section thicknesses.
(5) The requirement for holes, inserts and undercuts.

Another factor to be considered is whether a joining process would be more economic or more suitable for the form of the item concerned (see Chapter 8).

CASTING

Most metal products have at some stage in their manufacture been cast. *Casting* is the shaping of an object by pouring the liquid metal into a mould and then allowing it to solidify. The resulting shape may be that of the final manufactured object, or one that requires some machining, or an ingot which is then further processed by manipulative processes.

The mould used to form the shape into which the liquid metal is poured has to be designed in such a way that, however complicated the shape, the liquid metal flows easily and quickly to all parts. This has implications for the finished casting in that sharp corners and re-entrant sections have to be avoided and gradually tapered changes in sections used. Account has also to be taken of the fact that the dimensions of the finished casting will be less than those of the mould due to shrinkage occurring when the metal cools from the liquid state to room temperature. Moulds are generally made in two or more parts, which are clamped together while the liquid metal is poured into them, then separated, when the metal has solidified, to enable the finished casting to be extracted. Complex castings can be achieved by the use of moulds having a number of parts. Hollow castings or holes or cavities can be achieved by incorporating separate loose pieces inside the mould, known as cores.

There are a number of casting methods possible and the factors determining the choice of a particular method are:

(1) Size of casting required.
(2) The number of castings required.
(3) The cost per casting.
(4) Complexity of the casting.
(5) The mechanical properties required for the casting.
(6) The surface finish required.
(7) Dimensional accuracy required.
(8) The metal to be used.

Sand casting involves the making of a mould using a mixture of sand with clay, for the traditional moulding material. This is packed around a pattern of the casting, generally of a hard wood and larger than the required casting to allow for shrinkage. The mould is made in two or more parts so that the pattern can be extracted after the sand has been packed round it (*Figure 6.1*). Sand casting can be used

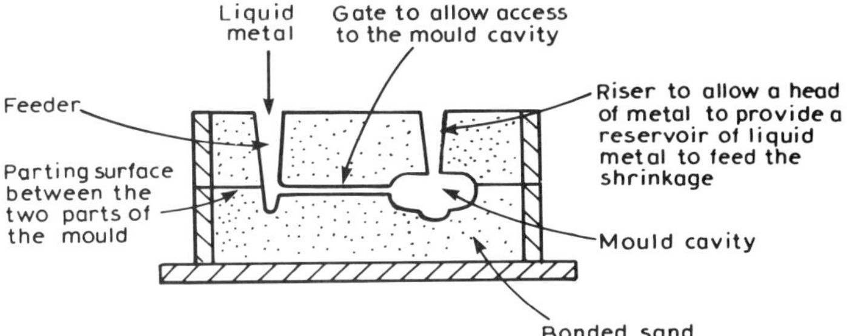

Figure 6.1 Sand casting

for a wide range of casting sizes and for small- or large-number production. It is the cheapest process for small-number production and a reasonably priced process for large-number production. Complex castings can be produced by this method. The mechanical properties, surface finish and dimensional accuracy of the casting are however limited. A wide range of alloys can be cast by this process.

Table 6.1 Typical manufacturer's information concerning sand casting with light alloys. (Courtesy of Sterling Metals Ltd).

A large pattern shop is operated for the manufacture and maintenance of patterns. Parts as large as 250 kg finished weight and as small as 10 g are produced regularly. On castings of average size a minimum wall thickness of 4 to 3 mm can be maintained. Typical tolerances for sand castings are ± 0.75 mm up to 150 mm. Allow a further 0.075 mm for each 25 mm over 150 mm. Where details are formed by cores or across a parting line additional tolerances will be required.

Die-casting involves the use of a metal mould. Two types of die-casting are used. *Gravity die casting* is similar to sand casting in that the metal mould has the liquid metal poured into it in a similar way to that adopted with a sand casting. The head of liquid metal in the feeder forces the metal into the various parts of the mould. With

Table 6.2 Typical manufacturer's information concerning die-casting. (Courtesy of Dynacast International Ltd).

The case for die-casting
Fastest of all casting processes.
High-speed, economical production.
Particularly suitable for very small parts.
High pressure die-casting results in fine-grain structure and good mechanical properties.
Complex, close tolerance parts made relatively easily.

The case for zinc
The most widely used alloy in die-casting.
Low melting point and excellent fluidity.
Easiest metal to cast.
Best alloy for small parts of complex shape and thin wall section.
Easily electroplated and mechanically polished.
Best mechanical properties after brass, which has poor castability.

pressure die casting the liquid metal is injected into the mould under pressure. This has the advantage that the metal can be forced into all parts of the mould cavity and thus very complex shapes with high dimensional accuracy can be produced. There are limitations to the size of the casting that can be produced by die-casting, that for pressure die-casting being smaller than that for gravity die-casting. The cost of the mould is high and thus the process is relatively uneconomic for small-number production. High-number production is necessary to spread the cost of the mould. These initial high costs may however be more than compensated for with large-number production by the reduction or complete elimination of machining or finishing costs. The mechanical properties, surface finish and dimensional accuracy of the casting are very good. The metals that can be used for this process are however restricted to the lower melting point metals and alloys, e.g. aluminium, copper, magnesium and zinc and their alloys.

Another method which is used to force the liquid metal into the various parts of the mould is known as *centrifugal casting*. The mould is rotated (*Figure 6.2*) and the forces resulting from this rotation force the metal against the sides of the mould. This method is used for simple geometrical shapes, e.g. large-diameter pipes. The method is not suitable for complex castings.

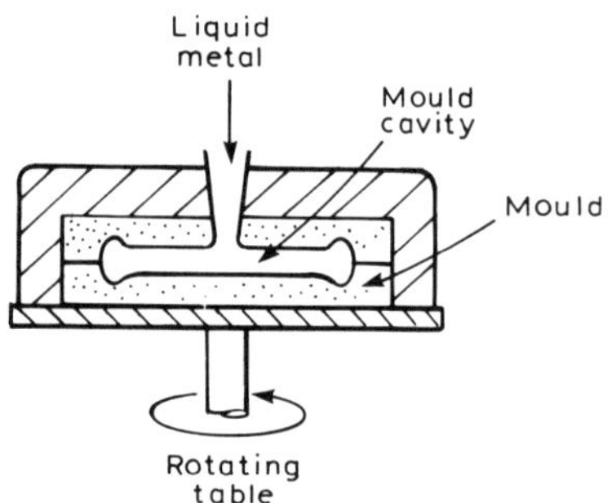

Figure 6.2 Centrifugal casting

Table 6.3 Typical manufacturer's information concerning centrifugal casting. (Courtesy of Holcroft Castings and Forgings Ltd).

The centrifugal casting technique has been developed to produce mainly circular forms such as gear wheel blanks, large bearing bushes and rings. Considerable grain refinement is achieved by the action of the spinning mould in combination with chilling, and the dense, uniform and homogeneous mass inherent in this type of casting gives increased strength with the greatest improvement at the periphery – an important consideration in the manufacture of worm wheels.

Foundry capacity
Centrifugal castings (wheel or rim castings)
Capacity 51-1879 mm diameter.
Maximum length dependent on diameter: up to 152 mm diameter, 155 mm maximum length; up to 1879 mm diameter, 254 mm maximum length.

Investment or *lost wax casting* is a process that can be used for metals that have to withstand very high temperatures, and so have high melting points, and for which high dimensional accuracy is required; aero engine blades are a typical product. The process is not restricted to high-melting-point metals but can also be used with low-melting-point metals. It is however the only casting method that can be used for the high-melting-point metals. Such metals cause rapid

Table 6.4 Typical manufacturer's information concerning investment casting. (Courtesy of Sterling Metals Ltd).

In order to obtain the maximum economic advantage, it is essential to design for the process. Thinner than normal wall sections are possible, and there are distinct dimensional tolerance advantages. Draw taper is not required. Within normal limits, overall size is not a limitation. Highly complex parts are possible, and in many cases a component may be produced as a one-piece casting rather than an assembly involving brazing, bolting, welding, etc. Critical dimensions should be agreed before commencement of an order, due to the fact that dimensional tolerances are affected by part complexity, section size, tool design and alloy.

die failures when used with die-casting. Modern investment casting uses metal moulds to produce wax patterns. The wax patterns are then coated with a ceramic paste. When this coated wax pattern is heated the ceramic hardens and the wax melts to give a ceramic mould. The liquid metal is then injected into this ceramic mould by pressure or the centrifugal process. After the mould has cooled the ceramic is broken away to release the casting. The size of castings that can be produced in this way is limited and it is an expensive process for large-number production. It is however relatively cheap for small-number production, particularly where high dimensional accuracy and good surface finish are required. It is suitable for complex castings (*Figure 6.3*).

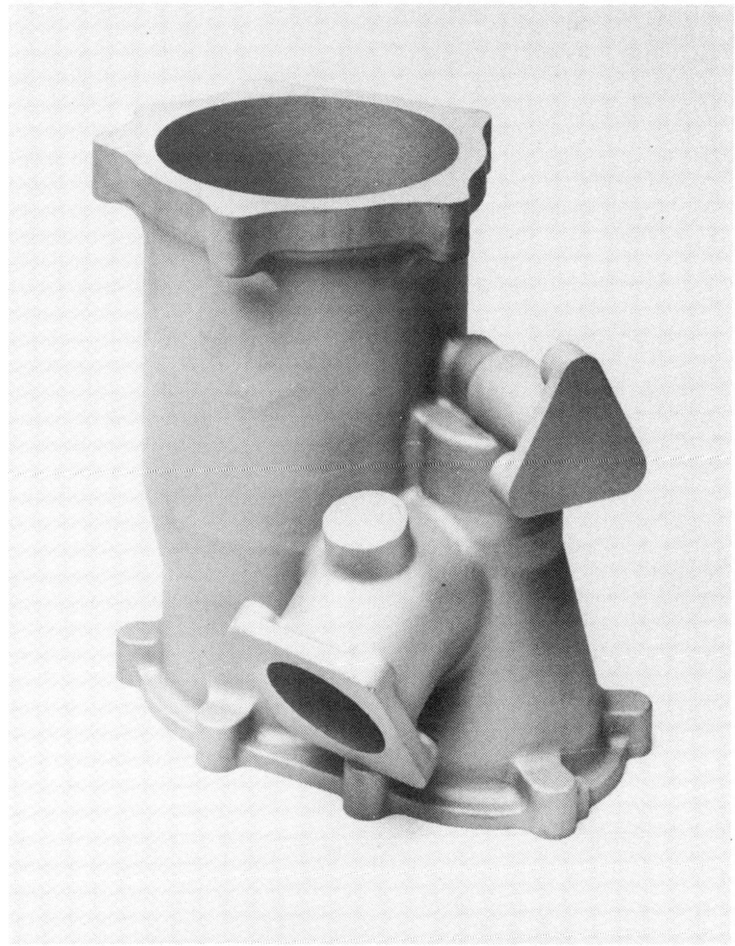

Figure 6.3 Pump body, investment cast in precipitation hardened stainless steel. (Courtesy of Sterling Metals Ltd)

MANIPULATIVE PROCESSES

Manipulative processes involve the shaping of a material by plastic deformation processes. Where the deformation is carried out at a temperature in excess of the recrystallisation temperature of the metal, the process is said to involve *hot working*. Plastic deformation at temperatures below the recrystallisation temperature is called *cold working*. The main hot-working processes are rolling, forging and extrusion. Cold-working processes are cold rolling, drawing, pressing, spinning and impact extrusion.

An increase in temperature at which a metal is worked means less energy is required to work the metal, i.e. the metal is more malleable. High temperatures can mean, however, surface scaling or damage occurring. The initial cast metal has coarse grains; hot working breaks the grains down to give a finer structure and thus better mechanical properties.

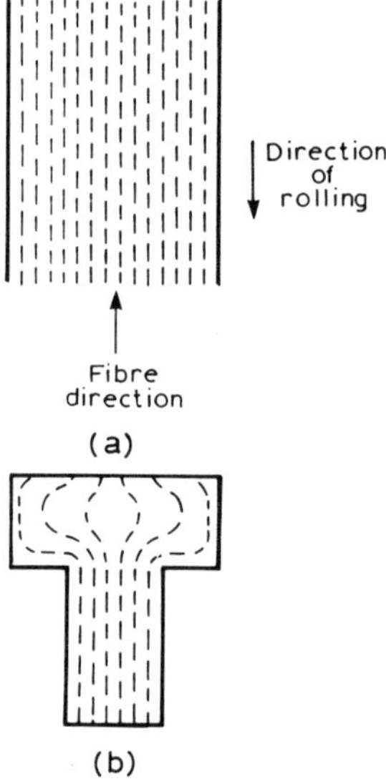

Figure 6.4 Fibre direction with (a) Rolling (b) Forging

In addition to the alloying elements present in a metal there are impurities derived from the fluxes and slags used in the melting operation. With the cast ingot these impurities are reasonably randomly distributed but with hot working they tend to become orientated as *fibres* in the direction of the working. Thus with rolled products the fibre lines tend to be in a direction parallel to the direction of rolling (*Figure 6.4a*). With a forging the work pattern, shown at (*b*), is more complex and so the fibre direction is more complex. The effect of the fibres having a specific direction is to give a corresponding *directionality* of mechanical properties. Thus although hot working improves the mechanical properties it does lead to the properties varying in different directions. The fibres can act as lines along which cracks can be propagated, so the design of a product should, as far as is possible, be such as to have the fibre direction, parallel to the tensile stresses and not at right angles to them.

Cold working involves the use of a greater amount of energy than a hot working process to obtain a particular amount of deformation. During cold working the crystal structure becomes broken up and distorted, leading to an increase in mechanical strength and hardness (*Figure 6.5*). Unlike hot working, cold working can give a clean, smooth surface finish.

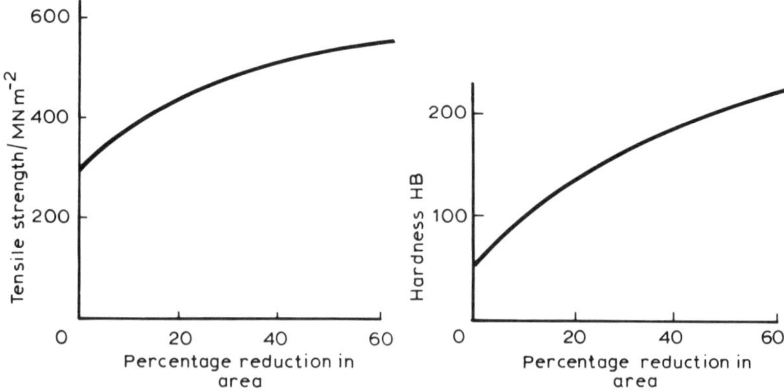

Figure 6.5 The effect of cold working on the tensile strength and hardness of a 85% Cu–15% Zn alloy

HOT-WORKING PROCESSES

The main hot-working processes are rolling, forging and extrusion. *Rolling* is the shaping of metal by the passing of the hot metal

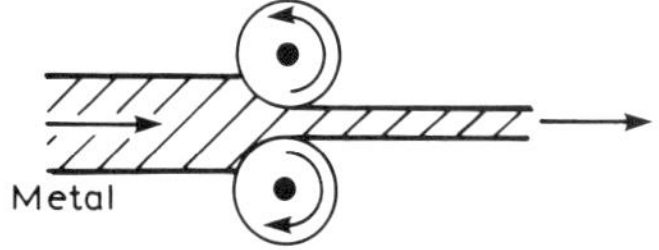

Figure 6.6 Basic principle of rolling

between rollers. *Forging* is the shaping of metal by a succession of hammer blows or application of pressure. *Extrusion* shapes metal by a hot ingot being forced, under pressure, through a die.

Rolling is a continuous process in which the metal is passed through the gap between a pair of rotating rollers (*Figure 6.6*). When cylindrical rollers are used, the product is in the form of bar or sheet, but profiled rollers can be used to produce contoured surfaces, e.g. structural beam sections (*Figure 6.7*). A wide variety of cross-sections

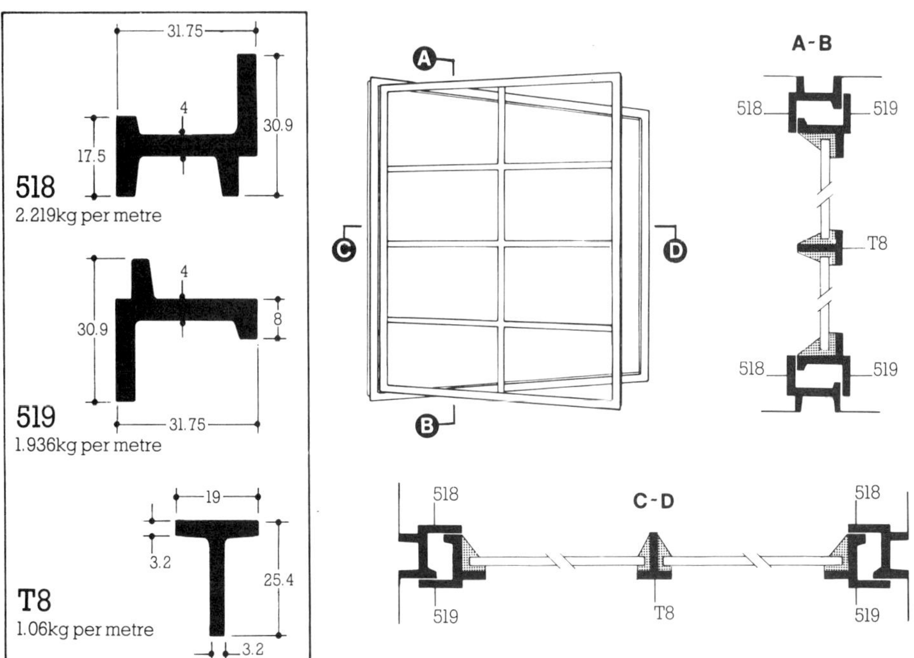

Figure 6.7 Rolled steel window sections. (Courtesy of Darlington & Simpson Rolling Mills Ltd)

can be produced. These are often used as the basis for structures, e.g. window frames, the rolled product being only trimmed to size and perhaps joined with other rolled shapes. In some instances the product may be further shaped by cold working to give the final product.

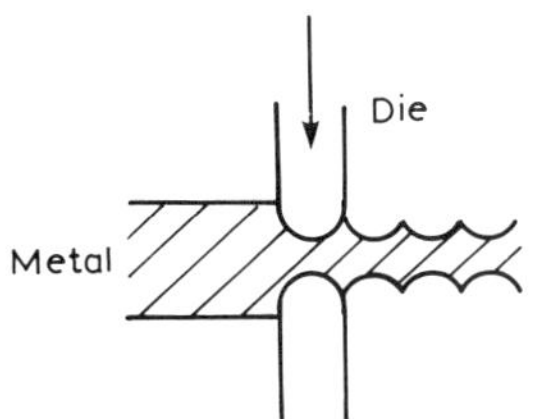

Figure 6.8 Open die forging

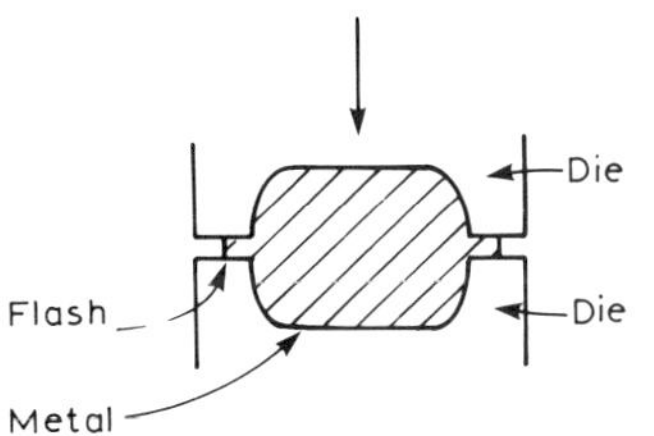

Figure 6.9 Closed die forging

With forging, the metal is squeezed between a pair of dies. *Heavy smith's forging* or *open die forging* involves the metal being hammered by a vertically moving tool against a stationary tool (*Figure 6.8*). This type of forging is like that once carried out by every village blacksmith, only now the ingot being hammered is likely to be considerably larger. The process is used nowadays, mainly for an initial rough shaping of an ingot before shaping with another process. *Closed die forging* involves the hot metal being squeezed between two shaped dies which effectively form a complete mould (*Figure 6.9*). The metal flows under the pressure into the die cavity. In order to completely fill the die cavity a small excess of metal is allowed and this is squeezed outwards to form a flash which is later trimmed away. *Drop forging* is one form of closed die forging and uses the impact of a hammer to cause the metal billet to flow and fill the die cavity.

Why extrusions?

Because virtually everything you design – in particular, anything where standard components are normally joined together – could be achieved by extrusions. More exactly. And, usually, more cheaply.

Because the extrusion is what you make it. Just what you want. With the metal where you want it. Your precise design – not an adaptation of standard components. Yet produced far more cheaply than many production methods can allow. So one part can do the job of several – eliminating costly machining and joining processes.

And, because the extrusion is aluminium it can be coloured and decorated so successfully that even a purely functional part can become an eyecatching design feature.

It's time **you** were using extrusion.

Extrusion stimulates imaginative design

More than any other production process, extrusion offers freedom to the designer; indeed, its versatility acts as a stimulus.

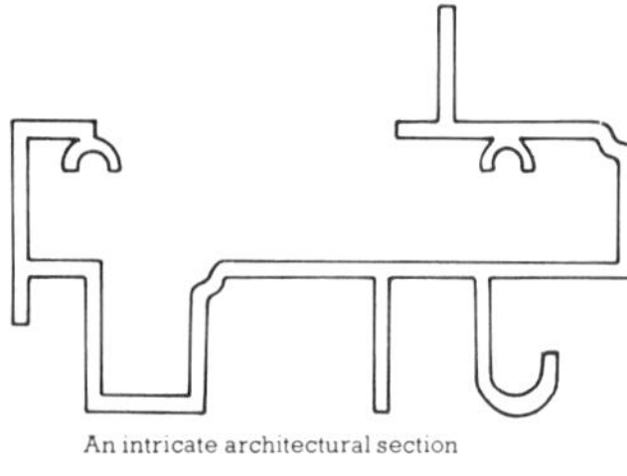

An intricate architectural section

Extrusion puts metal where you want it

The ease with which metal can be disposed to greatest advantage enables a section to be strengthened by the addition of bulbs or fillets and locally thickened to resist abrasion or provide bosses for screws.

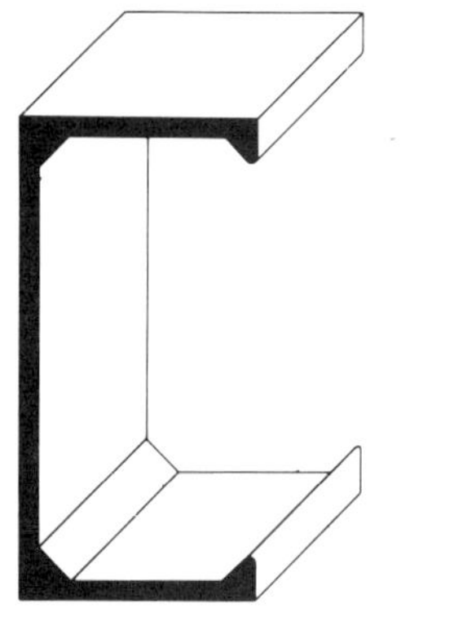

Bulbs incorporated in a standard channel section

Extrusion adds strength and stiffness

This is often possible when, for example, an extrusion is designed to replace a shape produced from metal strip.

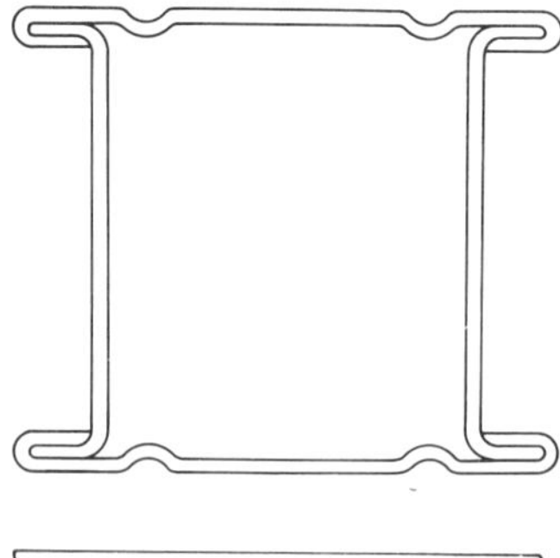

An extrusion replaces a shape made from roll-formed and crimped sheets

One extrusion can replace an assembly of parts

The result is usually a sounder, lower-cost and more attractive job.

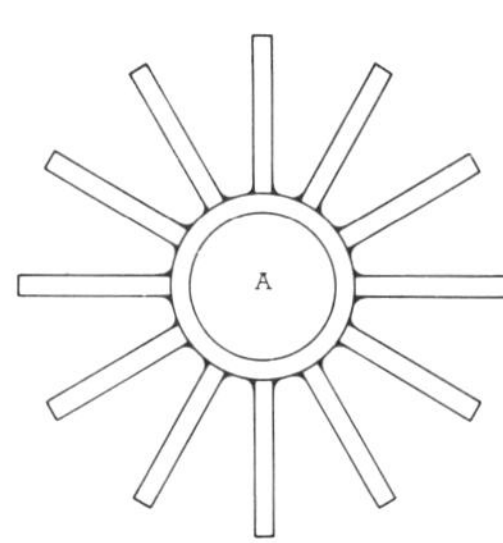

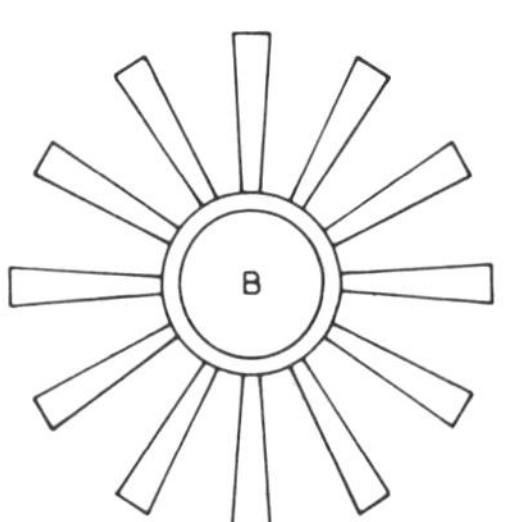

A finned tube produced by welding together numerous parts (A) is replaced by one extrusion (B)

Extrusion eliminates machining

Costs are saved by features such as integral slots and grooves that would otherwise have to be made by machining.

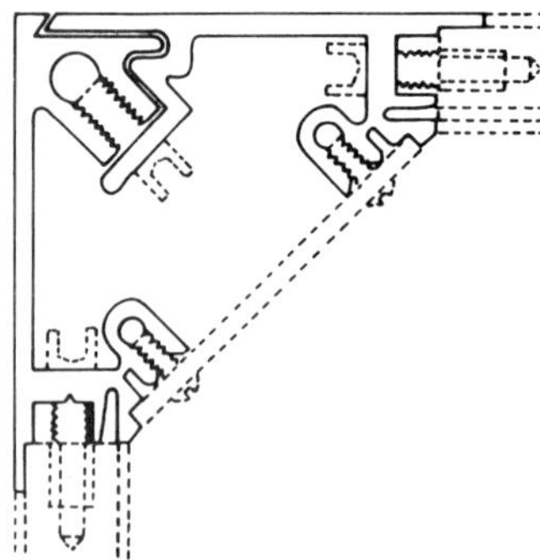

Corner post of a telephone kiosk, using two extrusions which incorporate slots for screws, eliminating machining

Extrusion tolerances allow close fits

This has considerable advantage, not least in the use of interlocking shapes for all types of fit.

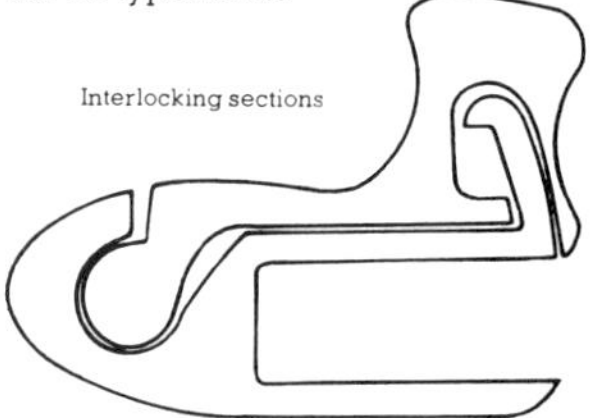

Interlocking sections

Why aluminium?

Because aluminium is strong yet light. Its strength to weight ratio is unmatched by any other commonly used material. Some of its alloys are even stronger than steel, but only one-third the weight. This light weight gives ease of handling and reduces labour and transport costs for you and your customers.

Because aluminium forms so easily it can be extruded into integral shapes of great sophistication and intricacy, with surprisingly low tooling costs . . . simple dies can cost under £150.

Because aluminium needs no protective coating and can be used in its natural state, yet colours far more successfully than conventional materials, whether by the unique colour anodising process or paint finishes.

Because aluminium won't warp or rot.

Because aluminium won't rust.

Because aluminium is non toxic.

Because aluminium has excellent thermal and electrical properties.

That's why aluminium . . .

Figure 6.11 (Courtesy of the Aluminium Extruders' Association)

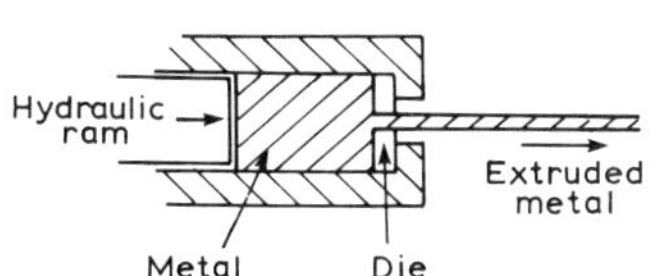

Figure 6.10 Extrusion

Closed die forging can be used to produce large numbers of components with high dimensional accuracy and better mechanical properties than would be produced by casting or machining. This is because the fibre direction can be arranged to give the greatest strength (*Figure 6.4*).

With hot extrusion the hot metal is forced, under pressure, to flow through a die, i.e. a shaped orifice (*Figure 6.10*). It is rather like squeezing toothpaste out of its tube. Quite complex sections can be extruded (*Figure 6.11*).

COLD WORKING PROCESSES

Cold rolling is the shaping of metal by passing it at normal temperature, between rollers (Table 6.5). Sheet and strip metal are often cold rolled as a cleaner, smoother finish to the metal surfaces is produced than if hot working is used. The process also gives a harder product. The aluminium foil used for wrapping sweets, such as chocolate, is an example of a cold rolled product. Cold rolling requires more energy than hot rolling.

Table 6.5 Typical manufacturer's information concerning cold rolled steel strip. (Courtesy of Arthur Lee and Sons Ltd).

Finishes

Bright hard rolled: General standard of finish as produced by cold rolling, which gives a smooth bright surface suitable for most requirements.

Bright annealed: Finally heat treated, maintaining a finish similar to bright hard rolled. Achieved by specialised and rigid atmospheric control in the annealing furnaces in which the heat treatment is carried out.

Plating or mirror finish: A superfine finish, provided in several plating grades for specific applications and requirements.

Matt finish: Produced by passing strip through specially prepared finishing rolls to give a dull but smooth surface. Very suitable for making parts which are to be lacquered, enamelled or painted. This strip has the advantage of retaining lubricants in the deep drawing process, thus facilitating this operation.

Coppered: Obtained by a process of electrolytic copper plating. Apart from its main application, it is also beneficial as a deep drawing lubricant, and as a base for articles to be plated, thereby saving production costs by the use of this partially prepared surface.

Patterned strip: Normally supplied in the annealed and pinch passed condition.

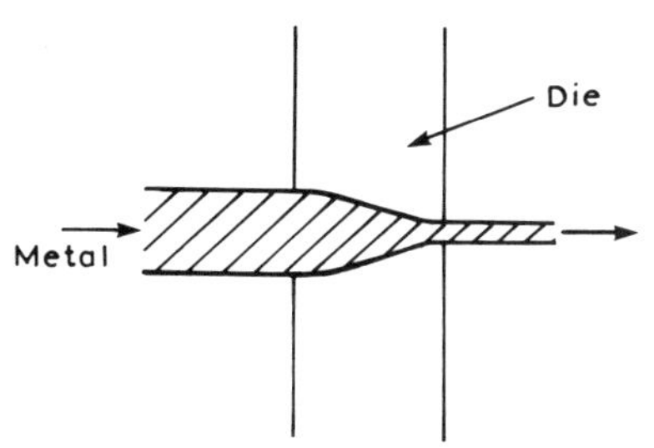

Figure 6.12 Drawing a wire

Drawing involves the pulling of metal through a die (*Figure 6.12*). Wire manufacture can involve a number of drawing stages in order that the initial material can be brought down to the required size. As cold working hardens a metal there may have to be annealing operations between the various drawing stages to soften the material for further drawing to take place.

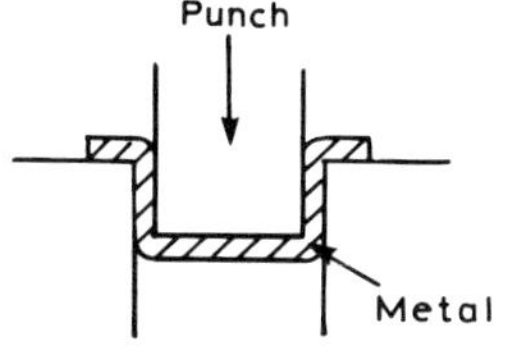

Figure 6.13 Deep drawing

With *deep drawing,* sheet metal is pushed through an aperture by a punch (*Figure 6.13*). The more ductile materials such as aluminium, brass and mild steel are used and the products are deep cup-shaped articles such as cartridge cases.

With deep drawing the sheet metal is not clamped round the edges and so is drawn into the die by the pressure from the punch. If the material is clamped round the edges then the process is known as *pressing* (*Figure 6.14*). Car body panels, kitchen pans and other cooking utensils are typical examples of the products obtained by pressing. Ductile materials are used.

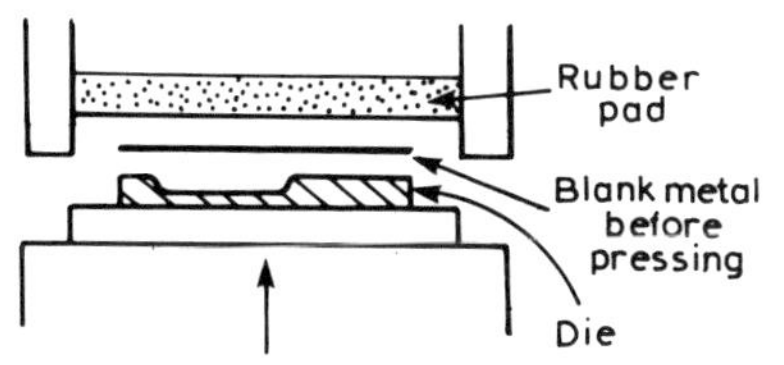

Figure 6.14 Pressing

Spinning is a process that can be used for the production of circular section objects. A circular blank of metal is rotated in a lathe type of machine and then pressure applied to deflect the blank into

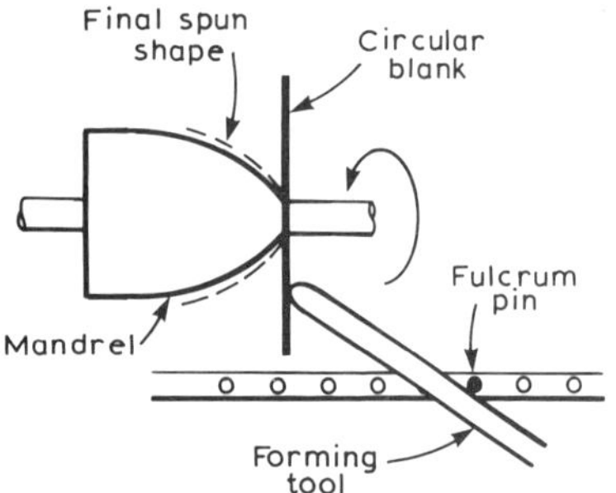

Figure 6.15 Spinning with a hand-held tool

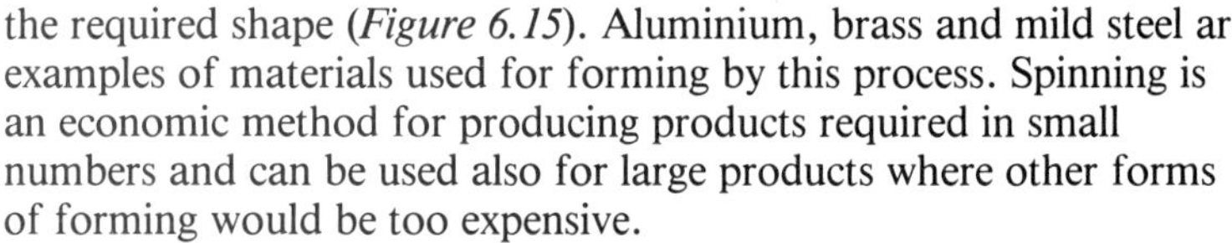
the required shape (*Figure 6.15*). Aluminium, brass and mild steel are examples of materials used for forming by this process. Spinning is an economic method for producing products required in small numbers and can be used also for large products where other forms of forming would be too expensive.

Explosive forming is used mainly for the forming of sheets of relatively large surface area on a comparatively small number production basis. An explosive charge is detonated under water and the resulting pressure wave used to press a metal blank against a die (*Figure 6.16*). Communication reflectors and contoured panels are examples of the products of this process.

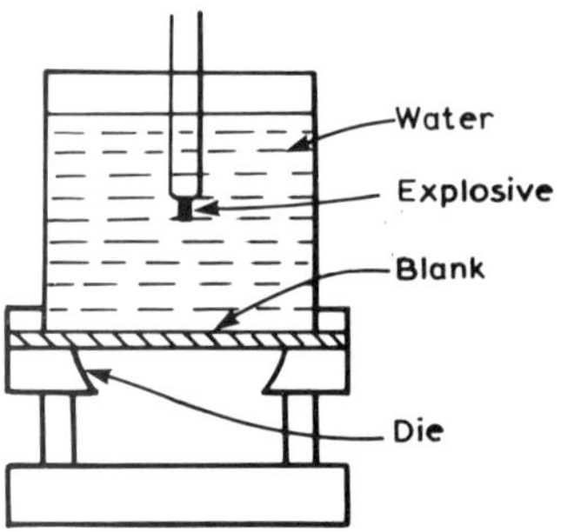

Figure 6.16 Explosive forming

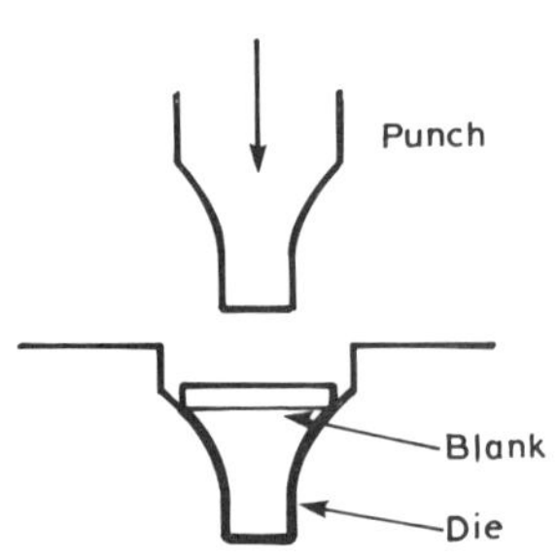

Figure 6.17 Impact extrusion

Impact extrusion is the process used for the production of rigid or collapsible tubes or cans, e.g. zinc dry battery cases and toothpaste tubes, in softer materials such as zinc, lead and aluminium. *Figure 6.17* shows the essential features of the process, a punch forcing a blank to flow into the die. The punch descends very rapidly and hence the term 'impact' for this form of extrusion.

POWDER TECHNIQUES

Shaped metal components can be produced from a metal powder. The process, called *sintering,* involves compacting the powder in a die, then heating to a temperature high enough to knit together the particles in the powder. Sintering of tungsten takes place at about 1600°C, considerably below the melting point of 3410°C. The sintering temperature for iron is about 1100°C.

Sintering is a useful method for the production of components from brittle materials like tungsten or composite materials. Cobalt-bonded tungsten carbide tools are produced in this way. The method is useful also for high melting point materials for which the forming processes involving melting become expensive. The degree of porosity of the metal product can be controlled during the process, which is thus useful for the production of porous bronze bearings. The bearings are soaked in oil before use and can then continue in service for a considerble period of time.

MACHINING

Machining is the removal of material from the workpiece, the block of material being machined, by the action of a tool. The tool moves relative to the workpiece and detaches thin layers of the unwanted material, known as 'chips'. *Figure 6.18* shows some of the main machining methods.

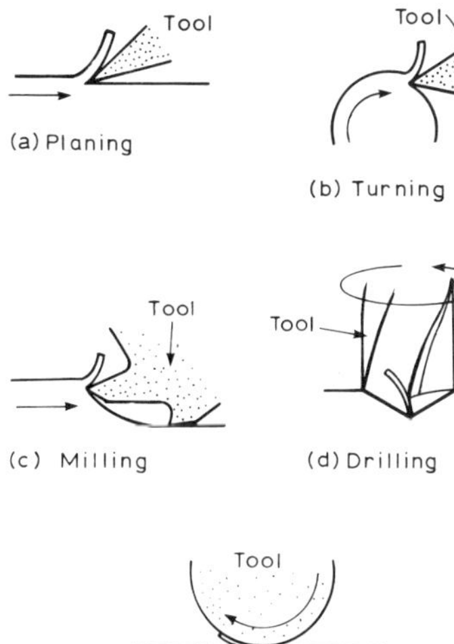

Figure 6.18 Machining methods (a) Planing (b) Turning (c) Milling (d) Drilling (e) Grinding

Machining is generally a secondary process, following a primary process such as casting or forging, and is used to produce the final shape to the required accuracy and surface finish. Machining results invariably in waste material being produced and thus any costing of a machining process has to allow for this. To keep the waste to a minimum the primary process should give a product as near the final required dimensions as possible, bearing in mind any need to remove material to give a good surface finish.

Machining involves using a tool that is of a harder material than that of the workpiece. Tools may be made of high speed steels, metal carbides or ceramics. Metal carbide and ceramic tools are made by sintering appropriate mixtures of powders (see previous section on powder techniques).

In machining, the cutting tool causes the workpiece material at the cutting edge to become highly stressed and subject to plastic deformation. The more ductile the material the greater the amount of plastic deformation and the more the material of the workpiece spreads along the tool face. The more this happens the greater the force needed to machine the material and so the greater the expenditure of energy in the machining process. A ductile material on machining gives rise to a continuous chip while a more brittle material leads to small discontinuous chips being produced. Less energy is needed for machining in this case.

The term *machinability* is used to describe the ease of machining. A material with good machinability will produce small chips, need low cutting forces and energy expenditure, be capable of being machined quickly and give a long tool life. Ductile and soft materials have poor machinability. A relative measure of machinability is given by a *machinability index*. It is only a rough guide to machinability, but the higher the index the better the machinability. These are some typical values:

Stainless steel	45
Wrought iron	50
Copper, $\frac{1}{4}$ hard rolled	60
Aluminium bronze	60
Cast steel	70
Free-cutting mild steel	100
Free-cutting α or β brass	200 to 400

The machinability of a metal can be improved if its ductility is decreased. Work hardening can make an improvement. Some multi-phase alloys have good machinability as the insoluble phase can provide discontinuities within the material and assist in breaking up a continuous chip to give small chips. The addition of a small amount of lead to a mild steel improves its machinability. For optimum machinability, the workpiece material must have low ductility and also low hardness.

Machining costs can very significantly increase the cost of, say, a casting. E. G. Donaldson in *Engineering* (March 1979):

> "The cost of machining castings can range from twice to over 20 times the cost of the casting itself. Consistent metal structure from batch to batch of melt, from casting to casting, and within any single casting allows the setting of machine rates to optimum, and allows these rates to be maintained. A casting of lower quality

results in slow-downs or downtime because of excessive tool wear, tool breakages, scrap, etc., leading to increased costs. The widespread introduction of numerically-controlled machine tools using preselected feed rates and cutting speeds means that there is less opportunity for operator intervention, for example in the case of a casting which is out of specification. Therefore a vital part of the economic machining of castings it to specify a suitable material from a reputable supplier for the basic casting."

Thus for economic machining, not only must a material have good machinability but the machinability must be consistent for all parts of the casting and for all the castings machined in the same way.

Table 6.6 Typical manufacturer's information concerning the machinability of metals. (Courtesy of Lee Bright Bars Ltd).

XLcut Free machining steels Why is XLcut better?

Sulphur: Sulphur, in the form of manganese sulphide particles, has the most significant effect on machinability. Ideally the particles should be thick oval shapes distributed evenly in the finished bar. They improve machinability by acting as a lubricant on the faces of the tool and reducing tool wear, modifying chip formation and reducing the shear forces produced during cutting and hence lowering power requirements.

Lead: The addition of lead substantially increases machinability over and above that attributable to sulphur alone. The lead occurs as envelopes around the manganese sulphides or as small discrete metallic particles. Further improvement in machinability is due to:
Additional lubrication of the tool tip which reduces tool temperature and hence tool wear, improvement in chip form and better surface finish.

Machinability index	
En 1A	100
XLcut	112
XLcutPb	130

GENERAL CHARACTERISTICS OF PROCESSES USED WITH METALS

Process	*Economic quantity*	*Materials (typical)*	*Optimum size*	*Min section possible*	*Holes possible*	*Inserts possible*
Sand casting	Small-large	No limit	1–100 kg	3 mm	Yes	Yes
Die casting-gravity	Large	Al, Cu, Mg, Zn alloys	1–50 kg	3 mm	Yes	Yes
Die casting-pressure	Large	Al, Cu, Mg, Zn alloys	50 g–5 kg	1 mm	Yes	Yes
Centrifugal casting	Large	No limit	30 mm–1 m dia	3 mm	Yes	Yes
Investment casting	Small-large	No limit	50 g–50 kg	1 mm	Yes	No
Closed-die forging	Large	No limit	3000 cm^3	3 mm	Yes	No
Hot extrusion	Large	No limit	500 mm dia	1 mm	—	No
Hot rolling	Large	No limit	—	—	No	No
Cold rolling	Large	No limit	—	—	No	No
Drawing	Small-large	Al, Cu, Zn mild steel	3 mm–6 m dia	0.1 mm	No	Yes
Spinning	One-off-large	Al, Cu, Zn mild steel	6 mm–4.5 m dia	0.1 mm	No	No
Impact extrusion	Large	Al, Pb, Zn, Mg, Sn	6 mm–100 mm dia	0.1 mm	—	No
Sintering	Large	Fe, W, Bronze	80 g–4 kg	0.5 mm	Yes	Yes
Machining	One-off-large	No limit	—	—	Yes	Yes

PROBLEMS

(1) Give examples of the types of product obtained by: (a) sand casting, (b) die casting, (c) centrifugal casting, (d) investment casting.

(2) Explain the meaning of the term 'directionality' and describe its effect on the properties of a product produced by hot working.

(3) What shaping processes might be used to produce a can, perhaps a Coca Cola can?

(4) Comment critically on *Figure 6.11.* Why extrusions? Why aluminium?

(5) Describe the different cold working processes and the types of products that they can produce.

(6) Explain how the machining of, say, steel can be improved by heat treatment or alloying additions.

(7) What type of materials are suitable for: (a) pressing, (b) impact extrusion?

(8) Compare the different methods of producing castings from the point of view of economic large number production.

(9) In an advertisement by Osborn Steel Extrusions the following claims are made:

"Osborn Steel extrusions might have been designed for designers of components. Countless shapes are available, a wider choice of materials and easier modifications are possible. Yet their overriding advantage is economy in labour, materials and machine time. They can replace machining, welding, rolling, forging and casting. Short runs make production more flexible. Stock control is simpler, you order just what you want when you want it."

Discuss the merits of the claims made.

(10) Deloro Stellite in an advertisement showing bottles of Haig whisky state:

"We help United glass turn out over 2500 Haig bottles an hour – a task in which high precision is even more important than high speed. (After all, the dimensional accuracy of any bottle is of paramount importance – especially when its contents are whisky.) Add to that the abrasive quality of molten glass – which creates havoc with conventional metals –and you can see why United Glass use Deloro Stellite wear resistant nickel based hardfacing alloys and castings for moulds and ancillary equipment."

What criteria determine the material used for moulds?

(11) Discuss critically the following extract, taken from an article 'Material cost trends favour metal forming' *Engineering* (October 1976).

"One of the intriguing questions when considering forming (particularly casting) as an alternative strategy to cutting, rather than as a complementary and contributory process, is, of course, whether machining can be completely eliminated. Not, of course, whether the metal cutting tool can be made redundant altogether – which, at this point in time, would be an utterly preposterous consideration – but to find the actual extent to which achievable capability of forming processes would provide complete fitness for purpose in a component. Metal cutting, as the normal method for the production of parts, dominates the thinking of parts designers. Lathes, mills, broaches and drilling machines are ubiquitous, and their very presence pre-empts the technique of production.''

(12) Explain the significance of the material referred to in the following extract from an advert, from the point of view of machining.

''More and more companies are discovering for themselves the advantages of changing to free chipping bar – and increasing their profits in the process.

Whether you use aluminium, brass or stainless steel, free chipping bar allows you to use higher speeds and feeds, thus reducing batch cycle time for the whole floor-to-floor operation.

By virtue of its chip-breaking properties, free chipping bar totally eliminates large spiral swarf, which, as you know, is not only bulky to store but can be positively dangerous to operators. Also there is far less chance of swarf build up at the tool point – and that can substantially increase tool life.''

Engineering, August 1978

10 ways not to have a component diecast

Mike Strowbridge of Alumasc Ltd tells you what to watch for if you are ordering diecastings

1 *Select your supplier on price alone.* A few pence saved at the quotation stage can prove a false economy if product defects show up after extensive machining, or failures occur once a component is in day-to-day use. Once the 'cheap' foundry has your tooling, they can manipulate price, quality and delivery to suit themselves. So the answer is look at the foundry's total capability at the beginning, not when it's too late.

2 *'Freeze' your design without detailed discussion with the foundry technical advisory centre.* This will ensure that you are unable to take the greatest advantages inherent in certain aspects of the diecasting processes. The most advanced foundry operations offer product-development services, applying value engineering and cost-reduction techniques at the conceptual stage of component design. This can be reflected in extended tool life and subsequent savings in tooling, material and component machining costs.

3 *Quality control is not possible in a foundry, so don't look for it in a supplier.* Far too often foundries have been the poor relation of production processes. Quality control *is* possible and the designer should look for it. Your foundry should have analytical control over all base alloys used, plus an approved quality-assurance scheme, backed up by real expertise in metallurgy.

4 *Design your product without giving consideration to the available diecasting techniques.* Every production technique imposes limitations. In the design of the product it is essential to observe the parameters imposed by the die design on the casting process. For instance, if minimum draft angles are ignored, this can cause problems on the foundry floor when removing the cast component from the die. So the moral is, talk to the foundry first.

Die cost, especially for large, complex dies, is relatively high and, combined with the high capital cost of diecasting equipment, requires that maximum utilisation and efficiency must be obtained to bring the component cost to a viable level.

5 *Stick to your preconceptions on which diecasting process to use, low-pressure, high-pressure or gravity, without taking expert advice.* Each diecasting method has its own advantages and disadvantages.

The low-pressure process permits complex castings to be produced economically, to close tolerances. Dimensional accuracy, surface finish, definition and pressure tightness characteristics are excellent. When compared to high-pressure diecasting, die costs are lower and shorter runs are an economic proposition. Highly stressed components can be produced in heat-treatable alloys.

Low-pressure diecastings

High-pressure diecasting comes into its own where long production runs justify the heavier initial cost of tooling and plant. The accuracy and high definition of high-pressure components and the use of cored holes instead of drillings, can eliminate many machining operations. The process allows thinner wall sections – between 1 and 3 mm – to be used while maintaining strength and rigidity in essential areas.

As one of the original methods of producing formed metal products, the gravity process has been known for thousands of years. Today the process still offers many attractions. It is particularly suitable for relatively short production runs in brass and aluminium, where the tools are simple in design. Tools can also be designed for continuous production.

6 *Select your diecasting supplier without considering whether or not he can offer back-up production services*. That will ensure production problems later on. Does it not make sense for the foundry to provide back-up services such as heat treatment, pressure testing, welding and machining? Perhaps you could solve your own production problems by buying machined components rather than raw castings.

7 *Don't consider who is going to be responsible for design and manufacture of tooling*. Any diecasting is only as good as the tooling. Some smaller foundries sub-contract their tooling design requirements. This only leads to split responsibilities and the possibility of varying standards of tool design. An in-house facility has the advantage of continuous, accumulating experience.

8 *Decide on what alloy you need for your diecast component without considering mechanical and physical properties and availability*.

Aluminium diecasting alloys

	Minimum mechanical properties			**Physical properties**		**General properties**				
	0.1% proof stress	Tensile strength	E	Density	Electrical conductivity % Cu	Gravity	Low pressure	High pressure	Corrosion resistance	Machinability
Specification	N/mm^2		%	g/cm^3						
LM2M	69.5	146.7	2.0	2.71	26	C	E	E	G	F
LM4M	69.5	154.4	2.0	2.74	32	G	G	G	G	G
LM5M	72.2	169.9	5.0	2.66	31	F	F	—	E	G
LM6M	61.8	185.3	7.0	2.66	37	E	E	G	E	P
LM8M	84.9	161.6	3.0	2.71	39	G	F	—	E	F
LM16	W9	247.2	2.0	2.71	36	G	G	G	G	G
	WP16	308.9	6.0							
*LM24M	108.2	278.0	3.0	2.71	24	—	—	E	G	F
LM25	72.2	139.0	2.5	2.68	39	E	G	G	E	F

Figures based on tests laid down in BS1490
* Pressure diecast P-Poor F-Fair G-Good E-Excellent

9 *Design sub-assemblies that can possibly be made cheaper with one diecasting*. The diecasting process allows complex shapes to be accurately reproduced. This makes it possible to replace a sub-assembly made up of a number of components with a single diecasting.

10 *Once an initial order is placed start to look around for competitive quotes which might save a penny or two per component*. You might think that it will keep the foundry on their toes. All it will do is sour relations built up over previous months during your new product's gestation period. The better relations are between designer and foundry, the better the service provided.

7 Forming processes with non-metallic materials

Objectives: At the end of this chapter you should be able to:
Distinguish the six main categories of polymer forming, i.e. extrusion, moulding, casting, calendering, forming and machining, and describe the types of product formed.
Identify the criteria which need to be considered before a choice of polymer-forming process can be made.
Describe the forming processes that are most commonly used with ceramics.

THE MAIN POLYMER-FORMING PROCESSES

Many polymer-forming processes are essentially two stage, the first stage being the production of the polymer in a powder, granule or sheet form and the second stage being the shaping of this material into the required shape. The first stage can involve the mixing with the polymer of suitable additives and even other polymers in order that the finished material should have the required properties. The additives may be in the form of solids, liquids or gases. Thus solid additives such as cork dust, paper pulp, chalk or carbon black are added to thermosetting polymers to reduce the brittleness of the material. Liquid additives may be used to improve the flow characteristics of the material during processing. Gas additives enable foamed or expanded materials to be produced.

Second-stage processes generally involve heating the powder, granule or sheet material until it softens, shaping the softened material to the required shape and then cooling it. The main types of process are:

(1) Extrusion
(2) Moulding
(3) Casting
(4) Calendering
(5) Forming
(6) Machining

The choice of process will depend on a number of factors, such as:

(1) The quantity of items required.
(2) The size of the items.
(3) The rate at which the items are to be produced.
(4) The requirements for holes, inserts, enclosed volumes, threads.
(5) The type of material being used.

Items required in a continuous length are generally extruded while items required in large quantities, particularly small items, are generally moulded, the injection moulding process being used. Casting and forming are relatively slow processes, unlike injection moulding which is much faster. Calendering is used for the production of sheet plastic. Forming involves the shaping of sheet plastic.

Table 7.1 Typical manufacturer's information concerning thermoplastics. (Courtesy of Bayer UK Ltd).

(a) *Durethan polyamide 6* (see *Figure 7.1*)
Form supplied: cylindrical granules
Colour range: available either natural or in opaque colours
Temperature performance: limit service temperature depending on type, 80–150°C
Conversion techniques: conversion from raw material: injection moulding, extrusion.
Conversion from semi-fabricated form: thermoforming, blowing, embossing.
Machining: drilling, turning, milling, sawing, cutting, tapping.
Jointing: non-detachable–bonding, welding, riveting.
detachable–mechanical and snap fits, screwed joints.
Decorating: painting, printing, metallising.

Predominant applications: electrical engineering, mechanical and precision engineering, automotive engineering, household appliances and domestic ware, building construction and furniture manufacture, packaging.

(b) *Makrolon polycarbonate* (see *Figure 7.2*)
Form supplied: cylindrical granules
Colour range: in all major colours, transparent, translucent or opaque, with excellent depth of shade.
Temperature performance: limit service temperature −150°C to 135°C unfilled, −150°C to 145°C filled.
Conversion techniques: conversion from raw material: injection moulding, extrusion, casting.
Conversion from semi-fabricated form: thermoforming, e.g. vacuum deep-drawing, pressurised air forming, blow forming, embossing.
Machining: sawing, drilling, reaming, milling, turning, shaping, filing, tapping, punching, cutting.
Jointing: permanent–bonding, welding, nailing, riveting.
detachable–mechanical and snap fits, screwed joints.
Decorating: painting, printing, metallising.
Predominant applications: electronic and electrical engineering, photographic equipment, lighting, optical equipment, mechanical and process engineering, precision engineering, safety and traffic equipment, office supplies, household and domestic ware.

Thermoplastic materials can be softened and resoftened indefinitely by the application of heat, provided the temperature is not so high as to cause decomposition. Because they flow readily with the application of heat they are particularly suitable for processing by extrusion and injection moulding; they can also be readily formed. Polyethylene, polyvinyl chloride, polystyrene, polyamide (nylon), polycarbonate, cellulose acetate and polytetrafluoroethylene are examples of thermoplastics.

Thermosetting materials undergo a chemical change when they are subject to heat which cannot be changed by further heating. Moulding and casting are processes often used with such materials. Typical thermosetting resins are phenol formaldehyde, urea formaldehyde, melamine formaldehyde, unsaturated polyesters and expoxides.

The properties of the material can be affected markedly by the inclusion of gas, to give foamed plastics. Thus, for example, the thermoplastic material polyvinyl chloride can be made into a flexible foam material for use, perhaps, as cushioning material. Polystyrene foam is used for thermal insulation. Thermosetting materials can also be used to give foamed plastics.

Another way of altering the properties of the material is to include reinforcement material within it, usually in the form of fibres. A thermoplastic or thermosetting material with, perhaps, glass fibres within it is likely to be stronger and stiffer than the plastic on its own. If the fibres are all aligned in one direction the strength and

Figure 7.1 (Courtesy of Bayer UK Limited)

Figure 7.2 (Courtesy of Bayer UK Limited)

stiffness in that direction will be markedly different from that in a direction at right angles to it. Glass fibre reinforced thermoplastic materials, e.g. nylon, is generally shaped by injection moulding.

Table 7.2 Typical manufacturer's information concerning a reinforced plastic. (Courtesy of Du Pont de Nemours Int., S.A.).

'Zytel' glass-reinforced nylon resins
In general, the addition of a defined percentage of glass fibres to a nylon resin enhances its strength, stiffness, toughness, fatigue endurance and creep resistance particularly at high temperatures. It also improves dimensional stability while retaining many of the property advantages offered by unreinforced nylons, such as excellent wear and solvent resistance, good electrical properties and a low coefficient of friction.

Glass-reinforced grades of 'Zytel' nylon resins, which are based on the 6.6 injection moulding grade reinforced with either 33% or 13% of short glass fibres, are designed specifically for applications in which shock loading is encountered and maximum toughness is required. Excellent resistance to impact is obtained. These resins have low moisture absorption and lower coefficients of linear thermal expansion than unreinforced types.

EXTRUSION

Extrusion involves the forcing of the molten polymer through a die. The process is comparable with the squeezing of toothpaste out of its tube. *Figure 7.3* shows the basic form of the extrusion process. The polymer is fed into a screw mechanism which takes the polymer through a heated zone and forces it out through the die. In the case

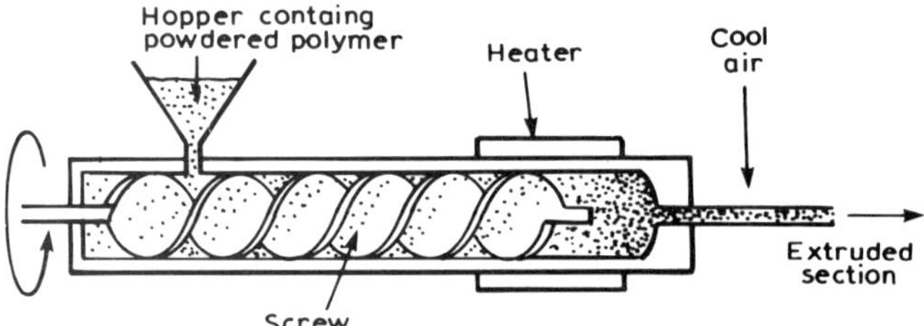

Figure 7.3 Extrusion

of an extruded product such as curtain rail the product is obtained by just cooling the extruded material. If thin film or sheet is the required product, a die may be used which gives an extruded cylinder of material. This cylinder while still hot is inflated by compressed air to give a sleeve of thin film. Another way of obtaining film or sheet is to use a slit die and cool the extruded product by allowing it to fall vertically into some cooling system.

The extrusion process can be used with most thermoplastics and yields continuous lengths of product. Intricate shapes can be produced and a high output rate is possible. Curtain rails, household guttering, polythene bags and film are examples of typical products.

Extrusion blow moulding is a process used widely for the production of hollow articles such as plastic bottles. Containers from as small as 10^{-6} m^3 to 2 m^3 can be produced. The process involves the extrusion of a hollow thick-walled tube which is then clamped in a mould. Pressure is applied to the inside of the tube, which inflates to fill the mould.

Table 7.3 Typical manufacturer's information concerning the extrusion of polymers. (Courtesy of Wragby Plastics Ltd).

Wragby is Britain's largest manufacturer of aerosol dip tubing, extruding more than 90 000 km (56 000 miles) per annum to tolerances of ±0.04 mm.

The Company has the capacity to extrude a multitude of different profiles for which it has an extremely high reputation within the custom extrusion field, the range of materials including p.v.c., polyethylene, polypropylene, ABS, and cellulose acetate.

Wragby is also the U.K.'s largest supplier of Dual In Line carrier tubes for the micro-chip industry.

MOULDING

A widely used process for thermoplastics is *injection moulding*. With this process the polymer is melted and then forced into a mould (*Figure 7.4*). High production rates can be achieved and complex shapes with inserts, threads, holes, etc. can be produced. The process is particularly useful for small components. Typical products are beer or milk bottle crates, toys, control knobs for electronic equipment, tool handles, pipe fittings.

Foam plastic components can be produced by this method. Inert gases are dissolved in the molten polymer. When the hot polymer cools the gases come out of solution and expand to form a cellular structure. A solid skin is produced where the molten plastic comes in contact initially with the cold mould surface.

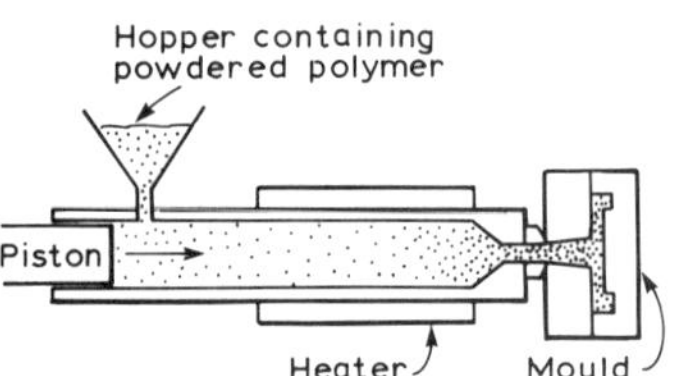

Figure 7.4 Injection moulding by a ram-fed machine. Screw-fed machines as in Figure 7.3 are also used

Table 7.4 Typical manufacturer's information concerning foam moulding. (Courtesy of British Industrial Plastics Ltd).

Structural foam moulding produces, in a single operation, a cellular core covered by an integral skin and gives a product which is three times as rigid as the same weight of solid plastic.

Structural foam competes with metal pressings and castings, wood and concrete, where rigid, strong, lightweight items are needed. It is suitable particularly for larger parts, can be nailed or stapled and has good screw retention properties. All conventional plastics finishing, painting and decorative techniques can be used.

The process is suitable for production of a wide range of items including furniture, toys, panels, containers, pallets and even small boats.

Although high density polyethylene, polypropylene, polystyrene and Noryl are the most popular, practically all thermoplastics materials can be moulded as structural foam.

Widely used processes for thermosetting plastics are *compression moulding* and *transfer moulding*. In compression moulding, the powdered polymer is compressed between the two parts of the mould and heated under this pressure (*Figure 7.5*). With transfer moulding, the powdered polymer is heated in a chamber before being transferred by a plunger into the mould (*Figure 7.6*).

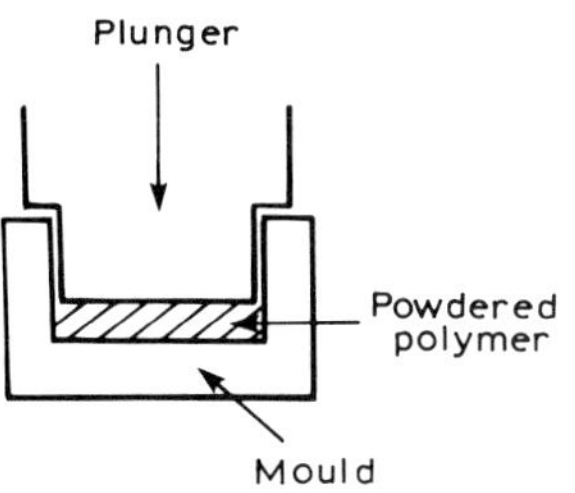

Figure 7.5 Compression moulding

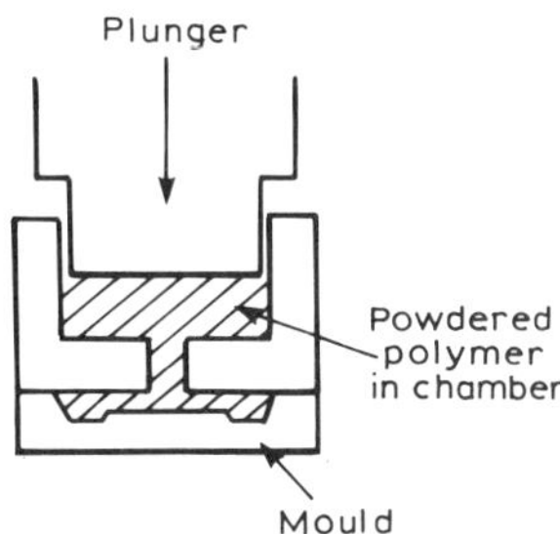

Figure 7.6 Transfer moulding

CASTING

One form of casting involves mixing substances of relatively short molecular chains, with any required additives, in a mould so that polymerisation, i.e. the production of long-chain molecules, occurs during solidification. The term *cold-setting* is used generally for such polymers. Such methods are used for encapsulating small electrical components.

Reinforcement of cold-setting polymers can be produced by incorporating fibres, e.g. glass fibres, within the polymer. *Hand lay-up techniques* can be used with fibre-glass mats and cold-setting polymers. The process involves coating a mould with a non-stick coating. A glass-fibre mat is then spread by hand over the surface of the mould and the liquid mixture is spread over the mat. Further layers of mat and mixture can be added until the required thickness of material has been produced. Then time is allowed for the reaction to proceed, i.e. the long chain molecule chains to be produced, and hardening occur. This type of technique is used by the many car owners when repairing rust or other damage to the metal bodywork of their cars. The technique has been used to produce entire car bodies.

Powder casting involves the melting of powdered polymer inside a heated mould. The mould is often rotated during this operation and the term *rotational moulding* used to describe the process. It is a very useful process for the production of hollow articles. This process does however have a slow rate of production. Powder casting can be used to coat surfaces with films of polymers, e.g. non-stick surfaces of cooking pans.

CALENDERING

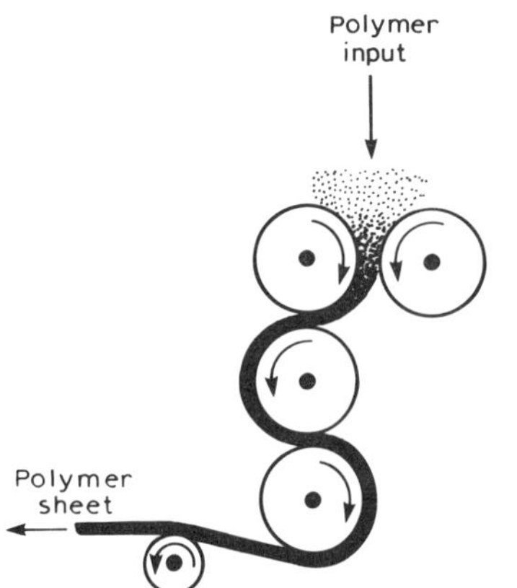

Figure 7.7 Calendering

Calendering is a process used for the production of continuous lengths of sheet thermoplastic, such as p.v.c. or polythene. The calender consists of essentially three or more heated rollers (*Figure 7.7*). The heated polymer is fed into the gap between the first pair of rollers, emerging as a sheet. The group of rollers determines the rate at which the sheet is produced, the thickness of the sheet and the surface finish.

FORMING

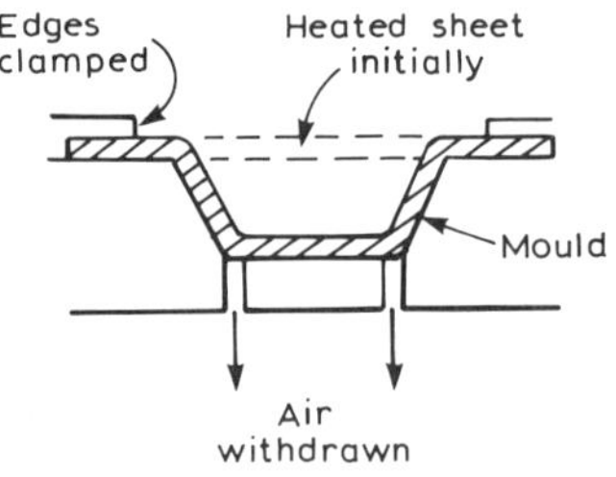

Figure 7.8 Vacuum forming

Forming processes are used to mould articles from sheet polymer. The heated sheet is pressed into or around a mould. The term *thermoforming* is often used to describe this type of process. The sheet may be pressed against the mould by the application of air under pressure to one side of the sheet, *pressure forming,* or by the production of a drop in pressure, a vacuum, between the sheet and the mould, *vacuum forming* (*Figure 7.8*).

Thermoforming can have a high output rate, but dimensional accuracy is not too good and holes, threads, etc. cannot be produced. The method can be used for the production of large shaped objects but not very small items. Enclosed hollow shapes cannot be produced.

MACHINING

While polymers can be machined by most of the methods commonly used with metals, the process used to shape the polymer often produces the finished article with no further need of machining or any other process. Polymers tend to have low melting points and thus correct machining conditions, which do not result in high temperatures being produced, are vital if the material is not to soften and deform. Some polymer materials are brittle and so present problems in machining, shock loadings having to be avoided if cracking is not to occur.

LAMINATES

A thin polymer coating can be applied to a substrate by extruding through a slit die and then pressing and simultaneously cooling the thin polymer film against the substrate. Other coating techniques involve depositing the thin layer from a solution of the polymer in a solvent or from the melt.

Coating a substrate with a number of polymer coatings can be achieved by bonding the films to each other and the substrate by

Table 7.5 Typical manufacturer's information concerning the machining of polymers. (Courtesy of Tufnol Ltd).

MATERIAL	AVAILABLE FORMS			MACHINING METHODS													
Thermosets (Laminates)	SHEET	ROD	TUBE	SAWING	TURNING	BORING	SCREW CUTTING	ROUTING	DRILLING	TAPPING	MILLING	SPINDLE MOULDING	PUNCHING	GUILLOTINING	SANDING	GRINDING	GEAR CUTTING
Phenolic and Epoxide—Paper	●	●	●	●	●	●	●	●	●	●	●	●	●	●	●	●	●
Phenolic and Epoxide—Fabric	●	●	●	●	●	●	●	●	●	●	●	●	●	●	●	●	●
Phenolic— Asbestos	●	●	●	●	●	●	●	●	●	●	●	●	●	●	●	●	
Phenolic & Epoxide—Glass Fabric	●			●	●	●	●	●	●	●	●		●	●	●	●	
Thermoplastics																	
Nylon 66	●	●		●	●	●	●	●	●	●	●	●	●	●			●
Nylatron GS	●	●	●	●	●	●	●	●	●	●	●	●	●	●			●
Nylatron GSM	●	●	●	●	●	●	●	●	●	●	●	●					●
Nylatron MC 901		●		●	●	●	●	●	●	●	●	●					●
Acetal	●	●		●	●	●	●	●	●	●	●	●					●

adhesives, solvent bonding or by the application of heat.

Paper coated with phenolic resin is widely used for electrical insulation purposes. Steel or aluminium is coated for decorative purposes, as well as reducing corrosion by chemicals in the atmosphere (see Chapter 5). Non-stick kitchen utensiles are plastic coated.

Table 7.6 Typical manufacturer's information concerning a laminated polymer. (Courtesy of Du Pont de Nemours Int. S.A.).

'Tedlar' P.V.F. film
High chemical stability and properties, unaffected by atmospheric conditions. Considered a real architectural innovation for the coating of building materials, this product can be bonded to or laminated with a great variety of materials such as galvanised steel, aluminium, vinyl wall cladding or polyester, giving them increased life and improved appearance.

GENERAL CHARACTERISTICS OF PROCESSES USED WITH POLYMERS

Process	*Production rate*	*Material type*	*Optimum size*	*Holes possible*	*Inserts possible*	*Enclosed hollow shapes*
Extrusion	Fast	Th, plastic Th, set	Few mm–1.8 m	—	No	No
Blowmoulding	Fast	Th, plastic	10^{-6} m^3–2 m^3	No	No	Yes
Injection moulding	Fast	Th, set	15 g–6 kg	Yes	Yes	No
Compression moulding	Fast	Th, set	Few mm–0.4 m	Yes	Yes	No
Transfer moulding	Fast	Th, set	Few mm–0.4 m	Yes	Yes	No
Lay-up techniques	Slow	Th, set and filler	0.01 m^2–400 m^2	Yes	Yes	Yes
Rotational moulding	Medium	Th, plastic	10^{-3} m^3–30 m^3	No	Yes	Yes
Thermoforming	Fast	Th, plastic	10^{-3} m^2–20 m^2	No	No	No

FORMING PROCESSES WITH CERAMICS

In general, the method used to form ceramics is to mould the material to the required shape and then heat it in order to develop the bonding between the particles in the material. As most ceramics are both hard and brittle the shape produced has generally to be the final shape as machining or cold working methods cannot be used.

A method used for the forming of clay shapes is *slip casting.* A suspension of clay in water is poured into a porous mould. Water is absorbed by the walls of the mould and so the suspension immediately adjacent to the mould walls turns into a soft solid. When a sufficient layer has built up, the remaining suspension is poured out, leaving a hollow clay object which is then removed from the mould and fired. This method is used for the production of wash basins and other sanitary ware.

The shaping of the block of wet clay on the potter's wheel is an example of *wet plastic forming,* the plastic clay mass being shaped by a tool before being fired. Another example of this type of forming is the extrusion of the plastic clay through dies, the extruded shape then being cut into appropriate lengths before being fired.

The *sintering* process, as described in Chapter 6 for metals, is used with ceramics. A version of this used with silicon involves compacting the silicon powder in an atmosphere of nitrogen at a temperature of about 1400°C. During the sintering process the silicon is converted into silicon nitride. This type of process is known as *reaction sintering* as it involves a chemical reaction as well as the sintering process.

PROBLEMS

(1) State a process that could be used for the production of the following products:

(a) plastic guttering for house roofs,
(b) plastic bags,
(c) the plastic tubing for ball-point pens,
(d) plastic rulers,
(e) a small, lightweight, plastic toy train,
(f) a plastic tea cup and saucer,
(g) the plastic body for a camera,
(h) a hollow plastic container for liquids,
(i) coating a metal surface with a coloured layer of polymer for decorative purposes,
(j) a plastic milk bottle.

(2) Describe the types of product produced by the following polymer processes:

(a) extrusion,
(b) injection moulding,
(c) calendering,
(d) thermoforming,
(e) casting.

(3) Describe the properties and some typical uses of foamed plastics.

(4) Spectacle frames can be moulded from thermoplastic cellulose esters. Explain the process that would have been followed.

(5) Why are plastic curtain rails made by extrusion rather than any other process?

(6) What types of process are used with thermosetting materials?

(7) The following is part of an advertisement from Wokingham Plastics Ltd in the December 1977 issue of *Engineering*. What types of product could be produced by the firm? Would there be any point, on the basis of the advertisement, in contacting them if you wanted someone to manufacture small toy railway engines?

"The case for vacuum forming is a case for Wokingham Plastics.

If it's plastic components you want – because they're tough, colourful, rust-free and inexpensive – contact Wokingham Plastics. We specialise. We have the necessary experience to form and fabricate industrial and domestic products. Such as fridge liners, car fascia panels, baths and showers, caravan interiors, light fittings. We work up to 2 m × 1 m, so virtually anything you need, we can make – in Polystyrene, ABS, PVC, Acrylic, CAB, Polycarbonate or Polypropylene."

Engineering, July 1976

10 ways to take advantage of plastics

The potential benefits of using plastics materials are considerable. Cost savings in high volume production, better chemical resistance, weight reduction, little or no finishing and easier assembly are all reasons why a designer might consider plastics rather than metals.

T E Jenner of Plasro Plastics shows the steps necessary for re-designing metal components to enable plastics alternatives to be satisfactorily produced. The principles outlined apply in the main to articles for injection moulding – the fabrication method most widely used.

1 The number of different materials and grades from which to choose is enormous and the designer is well advised to seek some specialist advice from raw material manufacturers and moulders. Material selection will depend on many factors including the following; temperature rating, chemical resistance, stress, product life, dimensional stability, tolerances, rigidity and assembly method. The behaviour of plastics materials under load is considerably different from that of metals, and a due allowance for the higher rates of creep must be made.

A brief list of plastics commonly used to replace certain metal components is given below.

Aluminium and zinc die castings	nylon – glass filled acetal resin – unfilled and glass filled polycarbonate – unfilled and glass filled modified polyphenylene oxide glass filled polyester – glass filled
Aluminium, brass and steel turnings, metal extrusions and pressings	as above, plus the unfilled grades of nylon, modified polyphenylene oxide and polyester
Gears and pulleys	acetal resin, nylon, polyester
Stainless steel components (chemical resisting)	polypropylene – unfilled, glass filled and talc filled polyphenylene sulphide
Aluminium and steel casework	polycarbonate, modified polyphenylene oxide, abs and polypropylene
Springs (not coil) and spring components	acetal resin

2 Even wall thickness will ensure minimum distortion, sinkage and loss of tolerance. These conditions are caused by the natural shrinkage of the material during cooling in the mould, where a greater shrinkage occurs in a heavier section than in a thinner one. Stiffness and strength need not necessarily be lost when eliminating heavy sections. Simple webbing can usually add stiffness where required. The actual wall thickness used will depend on the stress involved, the rigidity required and the minimum flow length of the material.

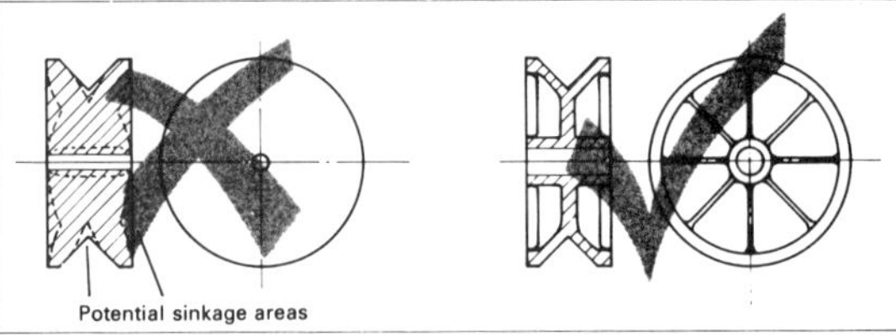

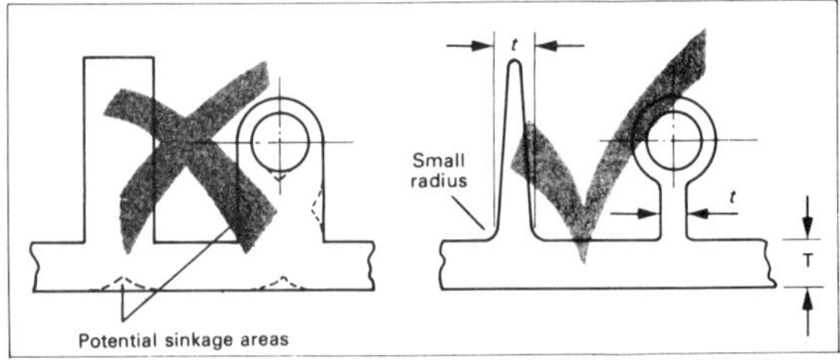

3 Correct web proportions should be used to reduce sinkage and distortion. The web thickness, t, should never exceed 0.6 of the wall thickness, T, and for high precision mouldings a web thickness of 0.4T is recommended.

4 Finish is an area of many possibilities, which should be considered as an integral part of design. If for instance a surface with deep grain effect is desired, an allowance of up to 7 deg draft will be needed on, say, case sides and this will considerably affect the product design. Apart from graining, spark erosion and vapour blasting of the mould give attractive finishes. The normal high polish finish can highlight flow marks in certain mate-

rials, nevertheless it is usually necessary for adequate moulding release. Of course, through the use of colours and the fact that the moulding is normally produced fully finished in a single operation, the designer can achieve appreciable cost savings. Secondary finishing techniques such as plating and hot foil printing further add to the designer's freedom.

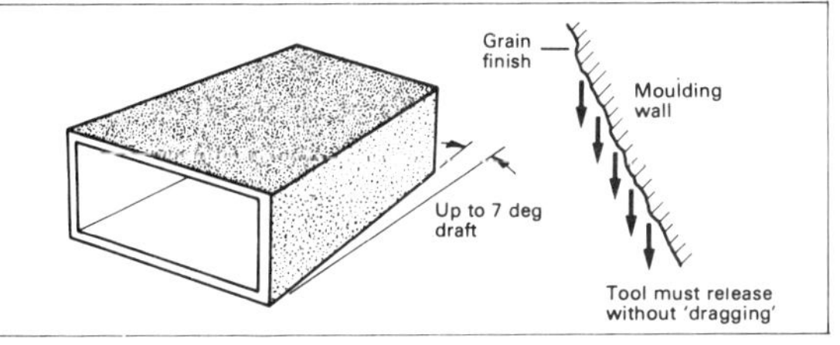

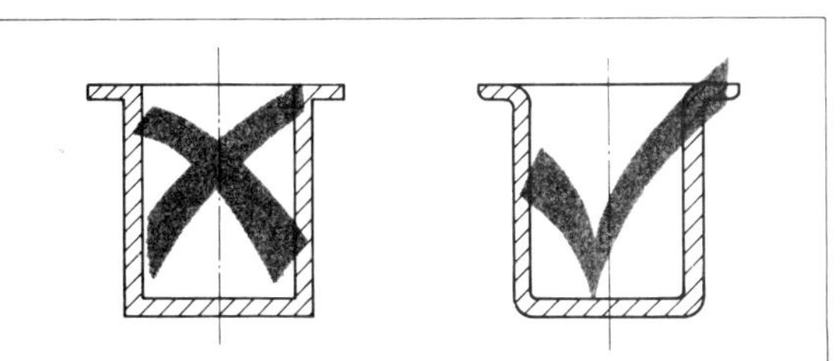

5 Radius corners, because like metals most plastics are notch sensitive. The presence of a notch, machining mark or sharp corner at changes of section or on threads, can induce a normally tough material to behave in a brittle manner. This is because the notch is in effect the start of a crack and very little energy is required to propagate the crack. A significant increase in impact strength can therefore be obtained by ensuring corners are radiused.

6 Assembly of components can very often be simplified by using the resilience of the plastics material, an example is the use of snap fits. There is also the possibility of cold heading, hot flaring of protruding bosses, welding, solvent bonding and adhesive bonding. The use of snap fits, however, is clean and efficient and obviates the need for additional materials and processes.

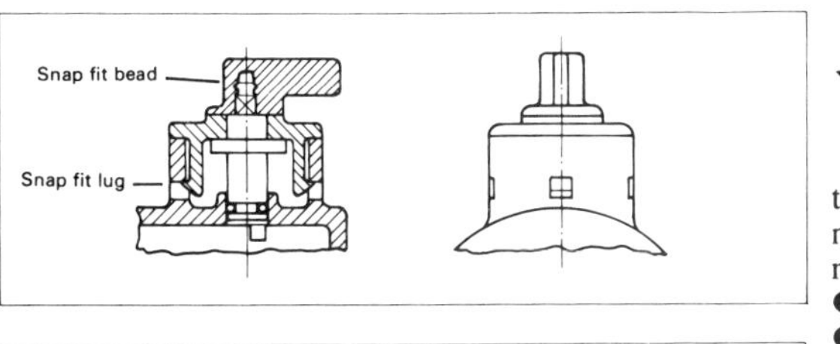

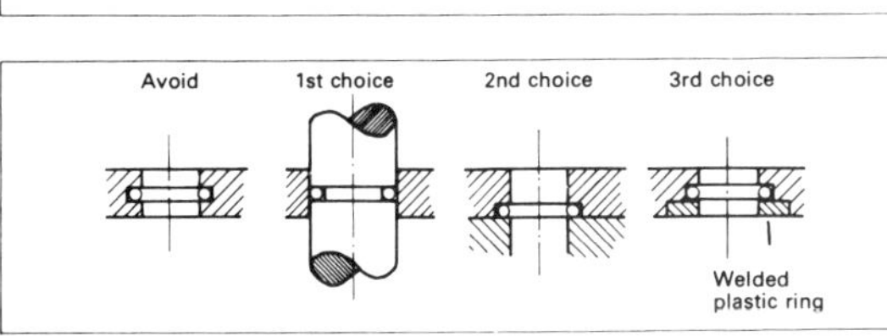

7 Design freedom is an advantage that should be fully exploited. Features that would be prohibitive in metals are produced simply in plastic mouldings, for example internal polygonal forms and integral springs. After-operations, however, should be avoided. Valve seats, keyways, metal inserts, integral hinges and O ring grooves should be moulded in. Deep undercuts should be designed out as these involve expensive collapsing core systems in the mould tool or after-machining operations; however, in resilient materials undercuts of up to $2^1/_2\%$ of component diameter can be ejected from the mould without difficulty.

8 Threads of any form either external or internal can normally be moulded into the component. The sketch illustrates an improved form for loaded conditions. Sometimes, however, it may be more convenient to form threads by other means – by tapping, use of self tapping screws or use of threaded metal inserts. All methods are possible. The most common methods of assembling metal inserts are – push fit, heated insert pressed in, ultra-sonic welding and moulded-in which is the strongest method.

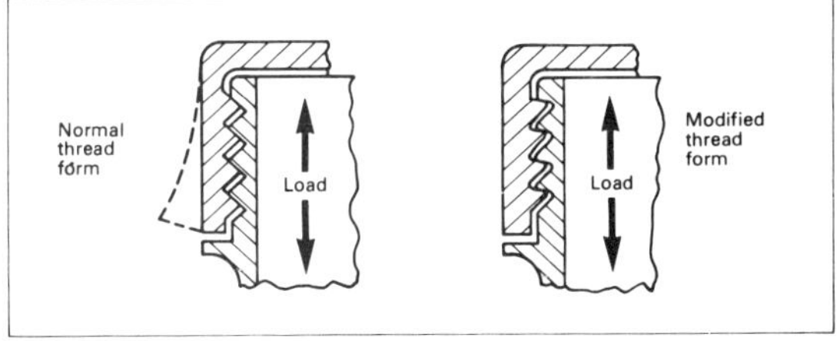

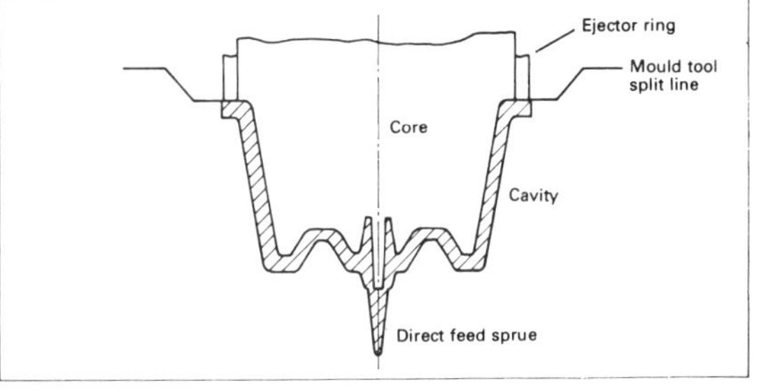

9 Draft is essential to facilitate release of the moulding from the tool, and an allowance must be incorporated in the design. This will vary according to the material used and whether it is on the tool core or cavity. As a guide about 1-1.5deg per side is normal.

10 Prototypes and models are useful for ensuring that all design aspects of the product are met, especially where unusual conditions are imposed. An allowance should be made for the fact that a machined prototype may behave differently from a moulded prototype or production item, usually for four reasons.

- Notches are created by the machining process
- The material selected for production is in a form (for example granules) unsuitable for machining and a less suitable alternative may have to be used for the model.
- To simplify machining, coring details and webs are usually left out.
- Material graining can adversley affect strength of an extrusion.

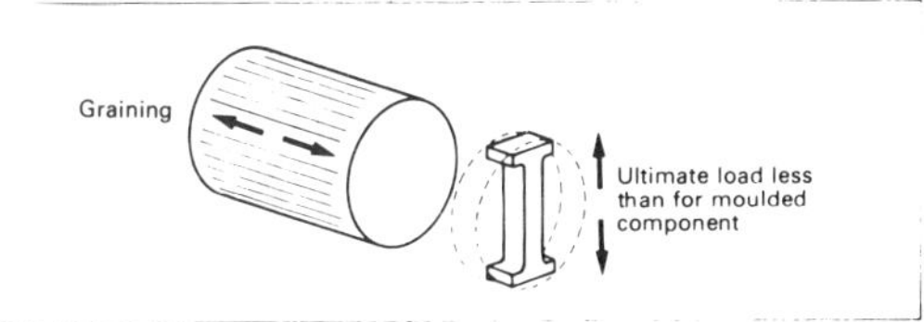

8 Joining materials

Objectives: At the end of this chapter you should be able to:
Describe the characteristics of the joining processes of adhesive bonding, soldering and brazing, welding and fastening systems.
Identify the criteria that need to be considered before a joining process can be chosen.

JOINING METHODS

The main joining processes can be summarised as:

(1) adhesive bonding,
(2) soldering and brazing,
(3) welding,
(4) fastening systems.

The factors that determine the joining process to be chosen are:

(1) the materials involved,
(2) the shape of the components being joined,
(3) whether a permanent or temporary joint is required,
(4) limitations imposed by the environment,
(5) cost.

ADHESIVE BONDING

Adhesives can be classified according to the type of chemical involved. These are:

(1) Natural adhesives.
(2) Elastomers.
(3) Thermoplastics.
(4) Thermosets.
(5) Two-polymer types.

Vegetable glues made from plant starches are examples of *natural adhesives,* which are used on postage stamps and envelopes. They set as a result of solvent evaporation. Such adhesives give bonds with poor strength which are very susceptible to fungal attack and are also weakened by moisture.

Elastomeric adhesives are based on synthetic rubbers; they also set as a result of solvent evaporation. Strong joints are not produced, there being low shear strength, and they are inclined to creep. These adhesives are used mainly for unstressed joints and flexible bonds with plastics and rubbers.

Thermoplastic adhesives include a number of different setting types. An important group of adhesives are those, such as polyamides, which are applied hot and solidify on cooling. They are used with metals, plastics, wood, leather, etc. and have wide application in rapid assembly work such as furniture assembly and the production of plastic film laminates. Acrylic acid diesters set when air is excluded, the reaction being one of a build-up of molecular chain length. Cyanoacrylates, the 'super-glues', have a similar type of setting action in the presence of moisture, the reaction taking place in seconds. Very useful for rapid assembly of small components. Other forms of thermoplastic adhesives set by solvent evaporation, e.g. polyvinyl acetate. In general, thermoplastic adhesives have low shear strength and under high loads are subject to creep, so they are used generally in assemblies subject to low stresses. They have poor to good resistance to water but are good with oil.

Thermosetting adhesives set as a result of a build-up of molecular chain length, to give rigid crosslinked matrices. Epoxy resins, one of the most widely used adhesives, are thermosets. Araldite is an example of an epoxy. They are two-part adhesives, in that setting only starts to occur when the two parts are brought together. They bond almost anything and give strong bonds which are resistant to water, oil and solvents. Phenolic resins are another example of thermosets. Heat and pressure are necessary for setting and they are used for bonding plywood. They have good strength and resistance to water, oil and solvents.

Combinations of thermoplastics or elastomers with thermosets are used to overcome some of the limitations present in the separate types of adhesive. Thermosets by themselves give brittle joints, but combined with a thermoplastic or elastomer a more flexible joint can be produced, with the high strength of the thermosetting bond. Phenolic resins with nitrile or neoprene rubbers have high shear strength, excellent peel strength, good resistance to water, oils and solvents and good creep properties. Setting occurs under pressure and at elevated temperature. Phenolic resins with polyvinyl acetate, a thermoplastic, require similar setting conditions and give similar bond strengths but with even better resistance to water, oils and solvents. These adhesives are used for bonding laminates and metals. Joints using them can be subject to high stresses and can often operate satisfactorily up to temperatures around 200°C.

For the maximum strength bonds to be realised with an adhesive, the joint should be designed to give the maximum area of bonding. Lap joints (*Figure 8.1*) are thus often used, the joints in *Figure 8.1b* and *c* giving greater stress uniformity across the joint than *Figure 8.1a.* There are many other forms of joint and the correct type should be chosen for the particular application concerned. For maximum strength there should also be a suitable adhesive, and the types of materials being bonded and the environmental conditions must be taken into account in this selection (Table 8.1). The conditions under which the adhesive are to cure need also to be taken into account, for example, there may be a need for the joint to be heated or be under pressure during the curing time, which may be a few seconds or many hours. Before application of the adhesive there may be a need to prepare the surfaces to which it is to be applied. Surfaces need to free from dirt, grease, moisture, etc. High-strength bonds may require the surfaces to be subject to abrasion with emery paper or be shot-blasted.

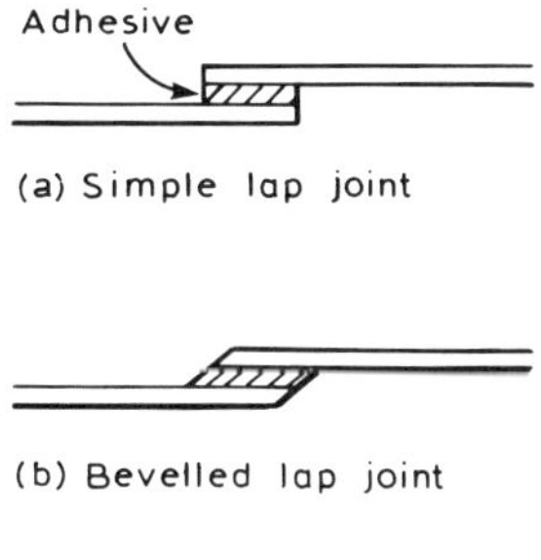

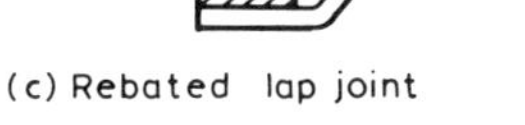

Figure 8.1 Adhesive joints

The use of adhesives to bond materials together can have certain advantages over other joining methods:

(1) Dissimilar materials can be joined.
(2) Joining can occur over large areas.
(3) Gives a uniform distribution of stress over the entire bonded area, with a minimisation of stress concentration.
(4) The bond is generally permanent.

Table 8.1 Chart showing complementary adhesives and adherents. (From Shields, J., *Adhesive Bonding,* The Design Council).

ADHESIVES / ADHERENDS	**Natural**	Animal glues	Starch	Dextrine	Casein	**Elastomers**	Acrylonitrile butadiene	Polychloroprene	Polyurethane	Silicone rubber	Polybutadiene	Natural rubber	Butyl	**Thermoplastics**	Cellulose nitrate	Polyvinyl alcohol	Polyvinyl acetate	Polyacrylate	Silicone resin	Cyanoacrylate	**Thermosets**	Phenolic formaldehyde	Urea formaldehyde	Resorcinol formaldehyde	Melamine formaldehyde	Polyesters (unsaturated)	Epoxy resins	Polyimides	Phenolic-vinyl formal	Phenolic-polyvinylacetal	Phenolic nitrile	Phenolic epoxy	**Inorganic**	Sodium silicate
Metals							X	X				X					X			X							X	X	X		X	X		
Glass, Ceramics		X						X							X		X			X							X			X		X		X
Wood					X							X			X		X					X	X	X	X		X							
Paper		X	X	X	X										X	X	X																	X
Leather		X					X	X				X			X																			
Textiles, Felt		X					X					X			X		X																	
Elastomers																																		
Polychloroprene (Neoprene)								X																										
Nitrile							X													X														
Natural							X					X								X									X					
Silicone										X																								
Butyl							X						X																					
Polyurethane							X	X	X																									
Thermoplastics																																		
Polyvinyl chloride (flexible)							X	X	X																									
Polyvinyl chloride (rigid)							X	X	X																		X							
Cellulose acetate									X						X					X														
Cellulose nitrate									X						X					X														
Ethyl cellulose															X					X							X				X			
Polyethylene (film)								X			X							X																
Polyethylene (rigid)																											X				X			
Polypropylene (film)								X			X							X																
Polypropylene (rigid)																											X				X			
Polycarbonate									X																		X							
Fluorocarbons												X							X			X					X							
Polystyrene									X											X							X							
Polyamides (nylon)								X														X		X			X				X			
Polyformaldehyde (acetals)									X											X						X	X							
Methyl pentene								X												X														
Thermosets																																		
Epoxy																				X		X		X			X							
Phenolic									X											X			X				X				X			
Polyester																										X	X							
Melamine								X	X																		X							
Polyethylene terephthalate							X	X											X							X								
Diallyl phthalate							X																			X	X							
Polyimide																											X	X						

Note: In general, any two adherends may be bonded together if the chart shows that they are compatible with the same adhesive

(5) Can be carried out at room temperature, or temperatures close to it, and so used where heat sensitive materials are involved.

(6) Leaves a smooth finish.

Among the disadvantages are:

(1) Optimum bond strength is usually not produced immediately; a curing time must be allowed.

(2) Many adhesives are flammable or toxic.

(3) The bond can be affected by environmental factors such as heat, cold and humidity.

SOLDERING AND BRAZING

With *soldering,* the joining agent is different from the two materials being joined but alloys locally with them. The joining agent, the solder, is heated together with the materials being joined until it melts and alloys with their surfaces. On cooling, the alloy solidifies forming a bond between the two materials. The joining process requires temperatures below 425°C and often below 300°C.

Solders are only weak structural materials when compared with the metals they are used to join; there is thus a need to ensure that the strength of the soldered joint does not rely on solder strength and is designed so that the materials interlock in some way (*Figure 8.2*).

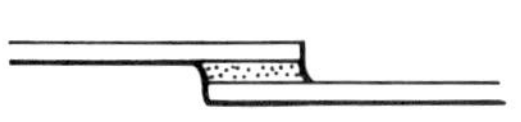

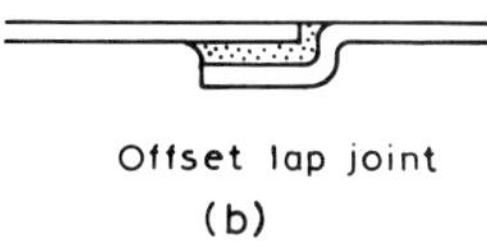

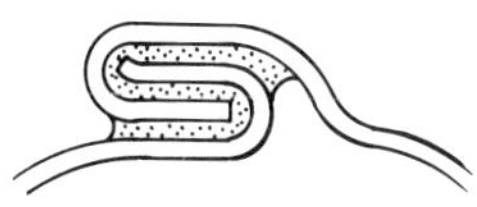

Figure 8.2 Solder joints

The hot solder must wet the metal surfaces being joined, which requires not only a suitable choice of solder material but clean surfaces, as unclean surfaces may make soldering impossible. Cleaning may involve abrasion of the surfaces as well as degreasing. Soldering flux is then applied to the surfaces. Fluxes, when heated, promote or accelerate the wetting of the surfaces by the solder. They remove oxide layers from both metal and solder and prevent them reforming during soldering. Fluxes are grouped in three categories: corrosive, intermediate and non-corrosive, and the least corrosive flux which will give a good joint should be used. After soldering the residues should bc removed; if left, they can result in corrosion of the metal surrounding the joint.

Solders are alloys of tin and other metals such as lead or antimony. The solder composition used depends on the metals being joined and the type of joint concerned. A 50% tin/50% lead solder could be used for joining sheet metal. For use at temperatures above 100°C a 95% tin/5% antimony solder may be used.

Brazing is a process similar to that of soldering but involves temperatures above 425°C. Brazing can be used with aluminium and its alloys, nickel and copper alloys, cast iron, steels, and many other less-used metals. Dissimilar metals can be joined. The term 'braze' comes from the use of brass as the substance used to make the joint. A 50% copper/50% zinc is used for general work, with a melting point of about 870°C. Other alloys, e.g. a copper/silver alloy, are used.

The procedure for brazing is similar to that for soldering, but the end result is a stronger joint than that given with solder.

WELDING

With brazing and soldering, the joint is effected by inserted a metal between the two metal surfaces being joined, the inserted metal having a lower melting point than that of the materials being joined. With *welding,* the joint is effected directly between the parts being joined by the application of heat or pressure. In *fusion welding* an external heat source is used to melt the interfaces of the joint materials and so cause the materials to fuse together. With *solid-state welding,* pressure is used to bring the two interfaces of the joint materials into intimate contact and so fuse the two materials together. Welding processes are capable of producing high-strength joints. The temperatures involved in producing the welds may, however, cause detrimental changes in the materials being joined. These may be local distortions due to uneven thermal expansion, residual stresses, or microstructural changes.

There are four main types of process used for fusion welding:

(1) Electric arcs

(2) Electrical resistance
(3) Radiation
(4) Thermochemical

With *electric arc welding* an arc is produced between the workpiece and an electrode. Temperatures of the order of 20 000 K are produced with currents between the electrode and workpiece of the order of 200 A. With *electrical resistance welding* the high temperatures are produced by passing an electric current across the interface of the joint and result from the passage of current through the electrical resistance of the joint. With *radiation welding* the high temperatures occur as the result of focusing a beam of electrons, in a vacuum or low pressure, on to the joint area. An alternative is to use a laser to focus a beam of radiant energy on to the joint. *Thermochemical welding* uses chemical reactions to generate the heat. One form of this uses a thermit reaction to produce liquid steel. Another form has oxygen and some fuel gas combining in a flame, oxygen and acetylene being very common.

Arc welding gives high quality welds, is a flexible method, and is low in cost. It is used for joints on bridges, piping, ships, etc. Electrical resistance welding can be used to give butt welds between two surfaces which butt up to each other, seam welds (a line of welded material between two sheets), or spot welds in which the weld occurs only in a small region, spots (*Figure 8.3*). Spot welding is used in the car industry for bodywork, seam welding is used in sheet metal fabrication. Thermit welding is used for the repair of iron and steel castings, railway lines, shafts, etc.

There are a number of forms of solid-state welding. *Pressure welding* involves a ductile material being pressed against a similar or dissimilar metal; aluminium and copper can be welded, cold, by this method. *Friction welding* involves sliding one material surface, under pressure, over the other. The friction breaks up any surface films and softens the surfaces by virtue of the rise in temperature produced by the friction. With *explosive welding* the two surfaces are impacted together by an explosive charge.

Solid-state welding is widely used for cladding sheets with a thin layer of some other metal. Aluminium alloy sheets are often clad with aluminium.

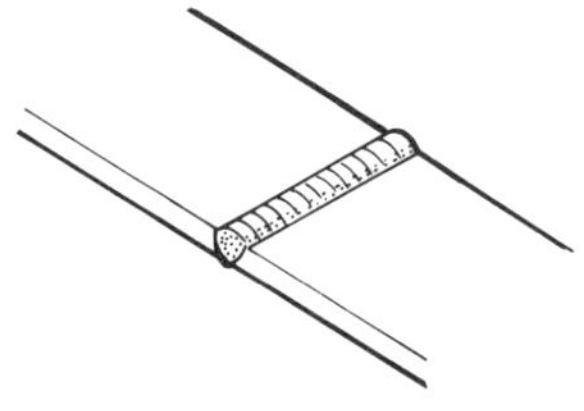

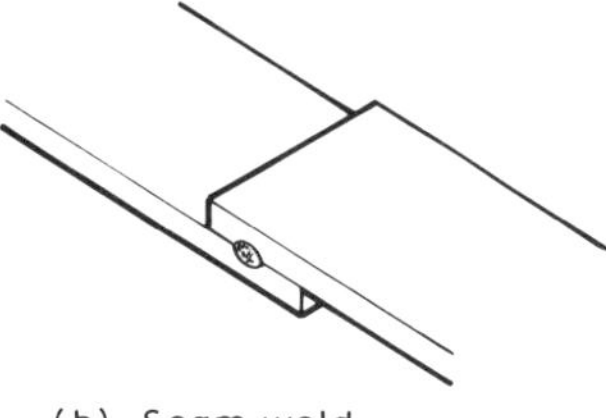

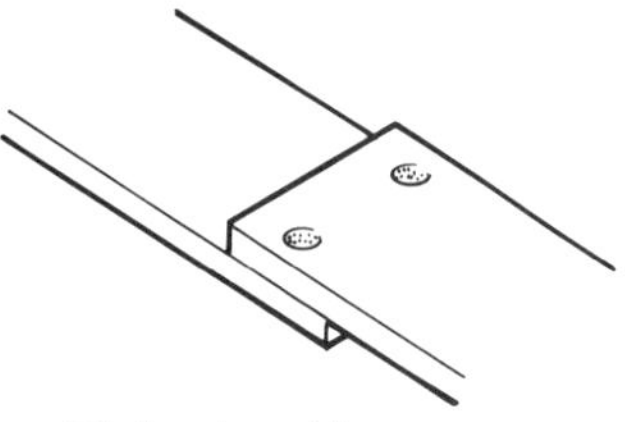

Figure 8.3 Examples of resistance-welded joints (a) Butt weld (b) Seam weld (c) Spot welds

FASTENING SYSTEMS

The choice of fastener will depend on a number of factors:

(1) Environmental, e.g. temperature, corrosive conditions.
(2) Nature of the external loading on the fastener, e.g. tension, compression, shear, cyclic, impact, and its magnitude.
(3) Life and service requirements, e.g. frequent assembly and disassembly.
(4) Design of the components being joined and types of material involved.
(5) Quantity of fasteners required and their cost.

Fasteners provide a clamping force between two pieces of material. A wide variation exists in types of fastener and the materials used for making them. The types of fastener available can be classified as threaded, non-threaded and special-purpose. Steel is probably the most common material used, although aluminium alloys, brass and nickel are among other metals used. Aluminium alloy fasteners have

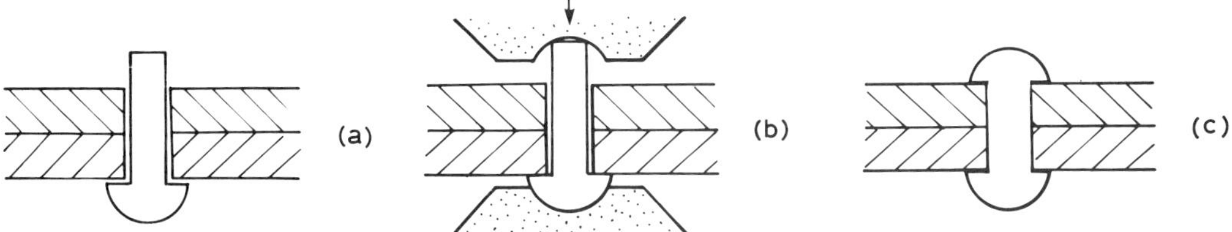

Figure 8.4 Typical stages in a riveting process

Figure 8.5 A cotter-pin joint

the advantage over steel of being much lighter, non-magnetic and more corrosion resistant. Nickel has the particular advantage of strength at high temperatures.

With a *threaded fastener,* the clamping force holding the two pieces of material together is produced by a torque being applied to the fastener and being maintained during the service life of the fastener. Bolts mated with nuts, and screws with threads in the material, are examples of threaded fasteners.

Nails, rivets and pins are examples of *non-threaded fasteners.* Nails are used extensively for making joints between pieces of wood. Rivets, however, are used for joining dissimilar or similar materials, both metallic and non-metallic. Both nails and rivets are low-cost fasteners designed for making joints which are intended to be permanent and not demounted. *Figure 8.4* shows the basic principle of riveting; the process is capable of use in automatic operations. *Figure 8.5* shows the principle of joints involving a pin. A cotter pin

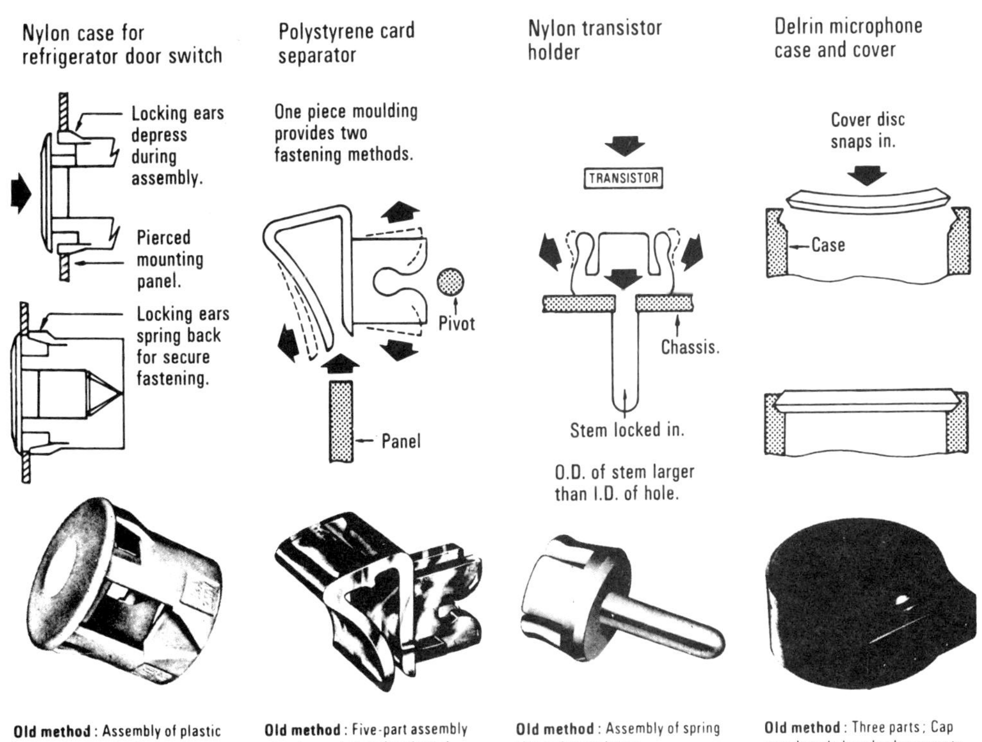

Figure 8.6 Examples of the types of fastening possible with plastic materials and the products possible with injection moulding. (Courtesy of Dynacast International Ltd).

is a simple means of fastening where some freedom of movement in the joint is required.

Figure 8.6 shows some examples of plastic non-threaded fasteners, which rely on the elastic properties of the plastic. Plastic fasteners are corrosion-resistant, light and generally cheaper than their metal counterparts.

PROBLEMS

(1) Discuss the merits of joining by means of adhesives and the limitations of the method.

(2) List the factors that determine the strength of an adhesive bonded joint.

(3) Suggest adhesives that could be used for bonding (see Table 8.1):

(a) metal to metal,
(b) natural rubber to a metal,
(c) nylon to nylon.

(4) Cans can be made by soldering sheet. What other methods could be used for cans and how do they compare with the soldering method? What method is used for Coca-Cola cans?

(5) Give examples of joints that have been made by:

(a) soldering,
(b) brazing,
(c) welding.

(6) Compare the types of joint that can be made by:

(a) threaded fasteners such as bolts and nuts,
(b) non-threaded fasteners such as cotter pins,
(c) non-threaded fasteners such as rivets.

(7) What type of jointing process can be used for:

(a) laminating a plastic sheet to chipboard,
(b) cladding aluminium alloy with aluminium,
(c) joining sheets of steel.

(8) In an article on Threaded Fastener Selection by J. M. Sharman (*Engineering* October 1975) the statement is made:

"The total life and reliability of a fastened joint depends not only on the initial strength of its component parts, but also on the compatibility of the materials from which they are made with the service environment."

(a) Explain the significance of this statement in relation to the choice of the material of the fastener and the material being fastened.

(b) What treatments might be used if the fastener is not made of the same material as that being fastened?

9 Choosing materials and processes

Objectives: At the end of this chapter you should be able to:
Analyse the material and process requirements for a component.

THE REQUIREMENTS

What functions does a component have to perform? What are the essential requirements of the component in order that it can fulfil those functions? These are two important questions that require answers before either the materials or the forming or fabrications processes can be established.

For instance, the functions of a domestic kitchen pan may be deemed to be: to hold liquid and allow it to be heated to temperatures of the order of 100°C. In order to do this, the container must not deform when heated to these temperatures. It must be a good conductor of heat. It must be leak proof. It must not ignite when in contact with a flame or hot electrical element. No doubt you can think of other essential requirements. There can also be other requirements which, though not essential, are desirable. For instance, the surface finish of the pan must look attractive.

From such an analysis of functions and requirements can come a list of the essential and desirable characteristics required of the material to be used for the object. The following are some of the common characteristics that could be required:

(1) Certain mechanical properties such as strength, hardness, stiffness, fatigue resistance.
(2) Certain behaviour in specific environments, e.g. corrosion free in a marine environment.
(3) Electrical or thermal properties, e.g. an electrical insulator or perhaps a good conductor of heat.
(4) Ability to be formed or fabricated to meet the requirements of the object's shape, dimensions, etc.
(5) A cost factor.

The choice of forming or fabrication process, or processes, will depend on a number of factors. The following are some possible factors:

(1) The quantity of items required and the required production rate.
(2) The dimensional accuracy required.
(3) The surface finish required.
(4) The size of the items, i.e. both overall size and section thickness.
(5) The requirement for holes, inserts and undercuts.
(6) The type of material being used.
(7) A cost factor.

In the case of the domestic kitchen pan, the requirement that the material be a good conductor of heat reduces the consideration to metals, particularly when taken together with the requirement that the material can be put in contact with a flame and contain hot liquids up to 100°C. A deep drawing process might be considered suitable with aluminium as the specified material. Being a cold working process this will give a good surface finish. But what about the pan handle? What material should be used for it? How should it be fastened to the pan? Perhaps it should not be fastened but produced in the same process as the pan.

PROBLEMS

(1) *A towel rail*
A simple towel rail consisting of a tube or rod held by clamps at each end to the wall is proposed. The towel rail is for use in the steamy atmosphere of the bathroom. The towel rails are to be mass produced and have to be cheap. The rail needs to withstand not only the weight of wet towels but also a person partially supporting their weight against the rail without any permanent bend being produced or indeed any significant bending under such loads.

(a) What are the crucial factors that determine the type of material that can be used?

(b) Suggest some possible materials.

(c) What determines whether a tube or rod is used? What determines the thickness of the tube or the diameter of the rod?

(d) What factors affect the cost of the finished component?

(e) Propose a possible specification for the material.

(2) *A simple spanner*
A mass production method is required for the production of spanners for sale as cheap items for the do-it-yourself enthusiast. The material used should be relatively cheap but able to withstand the uses to which the spanner will be put.

(a) What factors determine the type of material that can be used?

(b) What processing methods are feasible, bearing in mind the need to produce a cheap product? Consider the various merits and demerits of the various processes.

(c) What calculations would be needed to ascertain the thickness of material to be used?

(d) Specify a possible material and appropriate processing.

(e) How would you test the product to find out whether it was suitable for the design purpose.

(3) *An electric plug case*
Examine a 13 A mains electric plug. Such plugs are mass produced and cheap. They have to be designed so that they are cheap to produce and able to cope with the various terminals, fuse and connecting screws that are needed. They have to be electrically safe.

(a) What functions are required of the plug material?

(b) What tests would you need to run on a prototype material in order to check whether it is appropriate for a plug case?

(c) What materials would you suggest for the plug case? On what basis did you decide on the materials?

(d) What processes would you suggest for the production of the plug case? How are the processes determined by the material used? How are the material and process determined by the cost?

(4) *Motor car body*

For this problem you may like to consult 'The Open University, A second level course, An introduction to Materials Units 12 and 13, Car body'. Present day car bodies generally are made from mild steel with a carbon content of less than 0.1 per cent. The steel, in the form of thin sheet, is pressed into the required shapes.

(a) What functions are required of the car body which dictate the type of material used and the processes employed to shape the material? Consider the requirements to apply to mass produced cars selling at the lower end of the car market.

(b) Why is mild steel preferred over other materials?

(c) Would mild steel be an appropriate material to use for the skin of an aircraft? How does the material requirements of an aircraft differ from those of a car body?

(d) What would be the problems in the use of plastics for a car body?

(e) What are the problems associated with the use of mild steel for car bodies?

(5) *A street lamp support pole*

Consider the design and the materials employed for the street lamp support poles. Typical materials used are reinforced concrete, carbon steel and cast iron.

(a) What functions are required of the material used as a street lamp support pole?

(b) What processes might be used for the production of street lamp support poles?

(c) What part do you think cost plays in determining the material and process used?

(d) What part do you think ease of maintenance plays in determining the material used?

(e) Consider the production of a street lamp pole as an aluminium or steel extrusion. How would such materials behave in service? What would happen to the two materials if a car collided with the pole, i.e. a sudden impact?

(6) *A Coca-Cola container*

(a) Coca-Cola can be purchased in glass or plastic bottles and aluminium cans. What are the properties required of the material used for the container?

(b) What are the factors which lead to glass and aluminium being used for the containers?

(c) Consider the possibility of using a plastic container. What properties would you look for in the plastic? What factors would determine the feasibility of the plastic as the container material?

(d) What are the advantages and disadvantages of the glass and aluminium currently used for the containers? How would the plastic compare?

(7) *A suspension bridge*

One of the important features of any suspension bridge are the chains or cables that support the road deck. Some of the early suspension bridges used wrought iron for the chains, later bridges used steel. The suspension bridge over the river Severn near Bristol (England), completed in 1966, uses supporting cables consisting of 440 parallel high-tensile steel wires about 5 mm in diameter. The tensile strength of the cable material is about 1.4 GN/m^2.

(a) What are the properties required of the material used for the chains or cables used to support the road deck of a suspension bridge? What conditions are likely to be met in service and how will they affect the choice of material?

(b) What are the properties of steel that lend it to this use?

(c) The following extract is taken from the *New Scientist* of 15 June 1978.

"German bridge suffers rust in its stays.

The steel support cables on a German suspension bridge built in 1974 have become so corroded that they are to be replaced over the next eighteen months at a cost of £3 million. The 520 m Kohlbrand Bridge is part of a link between the port of Hamburg and a major motorway. The first signs of corrosion appeared two years ago and attempts were made to halt the process by coating the strands in a plastic based sealant. But within a year the cables near the road deck and the anchor points began to expand as the corrosion accelerated.

A survey just completed shows that rain had completely permeated the cable strands from top to bottom while salt applied to the road surface in winter also worked its way into the stays. Vibration from heavy vehicles using the bridge is also thought to have hastened the corrosion.

Since it is impossible to judge precisely how far the problem has advanced, Thyssen AG of Duisberg, which owns the original cable supplier, has decided to completely re-cable the whole structure. The 88 new cable stays will be heavily galvanised to provide better protection. The firm may also include vibration dampers in the cable anchor points. Thyssen will put up most of the costs, with the city of Hamburg also contributing."

(d) What were the factors contributing to the need to replace the cables?

(e) Explain the corrosion that was occurring and how, for the new cables, it is hoped to avoid this problem.

(f) What is the effect of the vibration?

(g) What tests would you suggest on the new cables before the cables are replaced?

(8) *A screwdriver*

The typical screwdriver is a steel driver in a plastic handle. Such items are mass produced and relatively cheap. They have to be able to withstand torques being applied to the driver without the driver becoming damaged or broken and also without rotating independently of the handle.

(a) What functions are required of (i) the driver material, (ii) the handle material?

(b) What types of material would be suitable for (i) the driver, (ii) the handle?

(c) What processes could be used to produce (i) the driver, (ii) the handle?

(d) How could the driver be joined suitably with the handle? What are the problems to be overcome in the joining process?

(e) What test procedures could be established to test the effectiveness of the product?

(9) *A toothbrush*

(a) What are the requirements of the materials used in a conventional toothbrush?

(b) What materials might be suitable for the handle and the brush parts?

(c) How might a toothbrush be produced?

(d) What material and process might be used to package a toothbrush so that it can be sold to the consumer in a prepackaged form?

(10) *Railway lines*

(a) What mechanical properties are required of the material used for railway lines?

(b) What environmental conditions must be taken into account in considering the possible materials for railway lines?

(c) What material or materials might be suitable for railway lines?

(d) What process might be used to produce railway lines?

(e) In Britain continuous railway lines are used; how might lengths of line material be joined together to give these continuous lengths?

(11) *Springs*

(a) Describe some typical functions of springs.

(b) What properties are required of the materials used for springs?

(c) For some specific application suggest a material that could be used. Justify the selection.

(12) *A car exhaust system*

Car exhaust systems appear in practice to have a relatively short life before corrosion damage becomes so serious that a replacement exhaust system is required.

(a) What conditions are exhaust systems subject to in use?

(b) What properties are required of the materials used for exhaust systems?

(c) What materials might be suitable for exhaust systems?

(d) What materials are generally used?

(13) *Flexible foam plastics as packing material*

Polyether foam is used widely as a packing material where goods need to be protected against shock and vibration. The material can be cut to the shape of the object being packaged and so enable the object to be completely surrounded by foam. Frequently however the object being packaged is not completely surrounded but held between pads or corner blocks of the foam.

(a) How is the foam able to protect the packaged object against shock?

(b) What properties are required of the foam? How rigid do you think the foam should be?

(c) What type of tests might be carried out to test the efficiency of a particular foam packaging material and the way it is used?

(14) *Coal chutes*

The power station at Cliff Quay, Ipswich, England, had the problem of selecting a material for chutes along which moist coal slid. The chutes were previously made of mild steel and required frequent repair and replacement due to the development of rust and erosion by the continual passage of the coal. They were replaced by aluminium chutes and a much better performance was then obtained.

(a) Why might aluminium be less damaged by the passage of the moist coal?

(b) What other materials would you have considered worth investigating?

(c) What tests might be used to compare materials for this purpose?

(15) *Small boats*

(a) What properties are required of the material used for the hull of a small boat?

(b) Materials that have been used for small boats hulls are wood, metals and composites such as glass-reinforced polyester resin. How do the properties of these materials compare with those required?

(c) What are the processes used to produce boat hulls from these materials?

(d) What factors do you consider determine the choice of material for a small boat hull?

10 Case studies

Objectives: At the end of this chapter you should be able to:
Establish the criteria that dictated the choice of material and process for particular components.
Extract and interpret information about materials and processes from case histories.

Answer the questions on each case study, after reading the corresponding article on the page indicated.

1. MATERIALS ASPECTS OF ENGINEERING DISASTERS

(Page 86)

J. G. Ball and R. D. Rawlings, *Engineering,* December 1978.

(1) What was considered to be the reason for the failure of the welded joints of the Kings Bridge?

(2) What procedure might have led to an earlier recognition of the defective welds?

(3) What information should have been disclosed by the Izod tests on the material used for the bridge?

(4) What procedure would you advocate for a laboratory carrying out impact tests in view of what occurred in the testing for the bridge material?

(5) Explain the significance of the information presented in the graph (*Figure 3*) in view of the failure that occurred of the Sea Gem tie-bars and legs.

(6) Why might the steel used for the Sea Gem have been appropriate when the rig was in the Gulf of Mexico and the Middle East but not appropriate in the North Sea?

(7) In the case of the pressure vessel, how were the investigators able to demonstrate that the vessel had not been stress-relieved?

(8) Why did the hip prosthesis fail?

(9) What factors limit the choice of material for use as a hip prosthesis?

(10) What factors have to be taken into account when selecting a material for use as a tie bar or leg of a North Sea oil rig?

2. MISUSE OF MATERIAL

(Page 92)

R. Ross, *Engineering,* December 1977.

(1) What are the two main problem areas, according to the author, that need to be taken into account by designers if failure is not to occur?

(2) Why is there a need to monitor the heat treatment process?

(3) What are the problems associated with electroplating?

(4) What are the problems associated with the painting process?

(5) What are the problems associated with welding and why are non-destructive tests not always adequate to show all defects in welds?

(6) What type of misuse can occur with the user of a component?

3. CORROSION BEHAVIOUR OF NI-RESIST CAST IRONS

(Page 94)

Inco Europe Limited.

(1) Explain the reference to the 'galvanic table' in the first paragraph of the article.

(2) What is meant by 'graphitisation'?

(3) What conclusion can you draw from *Figure 1*? How would you expect the graph to change if the seawater had (a) not been aerated, (b) been in a colder climate?

(4) Explain the significance of the 'comments' in Table 2.

(5) What is the effect on the corrosion/erosion results if the seawater is moving rather than still?

(6) What the advantages of using Ni-Resist cast irons for brine circulation pumps instead of grey cast iron?

(7) The Ni-resist cast iron type D-2 referred to in the article has these mechanical properties;

Tensile strength 370 N/mm^2
0.2% proof stress, min 210 N/mm^2
Elongation, min 7%
Impact energy, Charpy V-notch 13 J

What type of mechanical behaviour would you expect of this material?

4. GASKET MATERIALS

(Page 95)

R. A. Weeks, *Engineering,* February 1978.

(1) What characteristics are needed by materials suitable for use as gaskets?

(2) How do rubber and p.t.f.e. compare as gasket materials?

(3) What are the properties of cork that led to its use as a gasket material? What are the limitations for cork as a gasket material?

(4) What are the merits of spiral-wound metallic gaskets?

(5) Which gasket material or materials would be suitable for use as a gasket at temperatures of about (a) −150°C, (b) 400°C?

5. STAINLESS STEEL SAUCEPAN

(Page 97)

Advertisement by BSC Stainless.

(1) What are the problems that the material used for a saucepan has to cope with?

(2) What advantages might stainless steel have, over other metals, as the material for a saucepan?

(3) What process or processes might be used to produce a saucepan from stainless steel?

(4) Would the processes differ, or be cheaper, if a metal other than stainless steel were used, e.g. aluminium?

(5) What types of test procedure might be set up to compare the merits of saucepans made out of stainless steel and a variety of other metals? What performance would constitute a 'good' saucepan?

6. CORROSION WAR

(Page 98)

M. A. Jacobson, *Engineering,* April 1977.

(1) Compare the relative merits of the various external coatings used with a car.

(2) Why should internal coatings not be applied if there is already some surface rusting.

(3) Explain how sacrificial protection against corrosion can be used?

(4) What corrosion problems can occur when body repairs are made?

(5) What are the merits, and possible problems, with foam filling of car body sections?

7. STEEL-REINFORCED CONCRETE
(Page 101)

SHS the Builder, booklet produced by the British Steel Corporation, Tubes Division. (SHS stands for Structural Hollow Sections, in the extract, CHS stands for Circular Hollow Sections).

(1) Why is steel used to reinforce concrete?

(2) What are the requirements of the structural framing in a building?

(3) Why were the concrete columns in the Glasgow Royal Infirmary redevelopment changed to reinforced concrete using a circular hollow section?

(4) Explain the concern with fire protection and the effect of such concern on the choice of materials.

(5) How has the use of steel reinforcement changed the form of buildings? What would buildings look like if only concrete were used?

8. CERAMICS IN TURBINE DESIGN
(Page 103)

R. Jones, *Engineering,* September 1976.

(1) Why did the turbine blades made in the fifties using cermets fail?

(2) Why were ceramics considered for turbine blades instead of just using metals?

(3) How have the designers coped with the problem of the brittleness of ceramics?

(4) Explain the significance of the bend test histogram in *Figure 2.*

(5) What is the Weibull modulus?

(6) Explain the difference in processes involved in reaction sintering and hot-pressed sintering. How do the mechanical properties of the products formed by these two methods compare?

(7) Explain the reasons for the 'duo-density' approach to rotor production.

(8) Explain the testing procedure adopted for the ceramic rotor.

(9) Explain the ageing problem that occurs with ceramics.

9. PLASTICS AS ENGINEERING MATERIALS — NYLON 6
(Page 107)

F. J. Parker, *Engineering,* May 1979.

(1) Summarise the properties of nylon 6.

(2) What is the effect of glass fibre reinforcement on the properties of the nylon?

(3) What is the effect on the properties of the nylon of acrylic modification?

(4) What is the effect on the properties of water absorption by the nylon?

(5) What processes can be used to shape nylon 6?

(6) Explain the significance of the mould temperature during moulding on the properties of the product.

(7) Why was nylon 6 considered a suitable material for (a) the gear wheel, (b) the car fan, (c) the fluorescent-light-bulb end cap, (d) the industrial plug and socket?

(8) What are the problems associated with machining the nylon?

10. LOOSE OR TIGHT?
(Page 111)

S. H. Grylls, *Engineering,* June 1974.

(1) What criteria have to be met by the joining mechanism used for the blades of scissors? What mechanism is generally used?

(2) Explain why the author considers the joining mechanism for secateur blades in *Figure 3* is better than that in *Figure 8.*

(3) Describe the various methods used for joining a saucepan to its handle. Explain the advantages and disadvantages associated with each method.

(4) Compare the joining methods described in *Figures 19* and *20* for the attachment of the steering lever to stub axle at the front of a car.

(5) In view of the authors comments, examine the various methods used to hold two items together on items such as childrens' toys or machinery where one item has to be rotated relative to the other. Comment critically on the methods used.

11. MATERIALS PROBLEMS IN AUTOMOBILES

(Page 115)

The following is an extract from an article by M. Jacobson, entitled Automobile engineering — a spectator's view, *Engineering,* January 1978.

(1) What are the environmental conditions which the author considers rubber and elastomers can meet when used in cars?

(2) What problems occur with rubber and elastomers under the above conditions?

(3) What are the problems that are encountered with the rubber gaiters on rack and pinion steering and front wheel universal joints?

(4) What are the reasons, according to the author, for the short life of water hoses? What type of properties are required of the material used for water hoses if they are to have a long life?

(5) What were the problems met when nylon was first used for knuckle joints? Why was nylon used?

Materials aspects of engineering disasters

From time to time the public is made aware of the inadequacies of engineering design, errors in construction or failures of materials by the publicity given to major engineering disasters. Alongside the few occurrences that have nationwide, or even worldwide, repercussions, there are also many instances of minor failures from which lessons of equal technical importance can and should be learned. For many of the failures, both minor and major, the properties of the materials used contributed to, or were the prime cause of, the failure. Even in cases where the failure cannot be directly attributed to the materials used, a thorough examination, e.g. metallographic, chemical analysis, fractography, can indicate the main factors responsible for the failure. Little can be learned from a general review of engineering failures and it is necessary to look at some of the detail to appreciate how things go wrong and what precautions can be taken to ensure the maximum safety in future designs. This article by Professor J G Ball and R D Rawlings* concentrates on the role played by materials in a small selection of instances of failure

First, to consider the partial collapse of the Kings Bridge in Melbourne as the result of the fracture of seven welded girders as a heavy lorry crossed the bridge (Fig. 1). The bridge had been constructed from a low-alloy steel to BS.968.1941 (high-tensile, fusion-welding quality, structural steel) with additional specifications for Izod and weldability tests which were intended to protect the bridge from brittle fracture.

Starting on the composition of the steel, we find that the analysis of the heats from which the failed girders were constructed were at the maximum permissible limits of composition for BS.968.1941.

	Composition by percentage weight			
	C	Mn	Cr	Mn + Cr
BS.968.1941	0.23 max.	1.8 max.	1.0 max.	2.0 max.
Heat 55	0.21	1.70	0.23	1.93
Heat 56	0.23	1.58	0.24	1.82
Girder (From Heat 55)	0.25	1.75	0.25	2.00
Girder (From Heat 56	0.26	1.70	0.25	1.95

Table 1: Comparison of heat and girder analyses with composition as specified by BS.968.1941

However, subsequent to the collapse the compositions of the failed girders were found to be outside the specified limits (Table 1). A lack of appreciation that a sample taken from a heat gives only an indication of the composition of a particular component made from that heat, together with ignorance of the influence of chemical composition on weldability, meant that no questions were raised about the suitability of the material for making crack-free welds. Certainly, the high C and

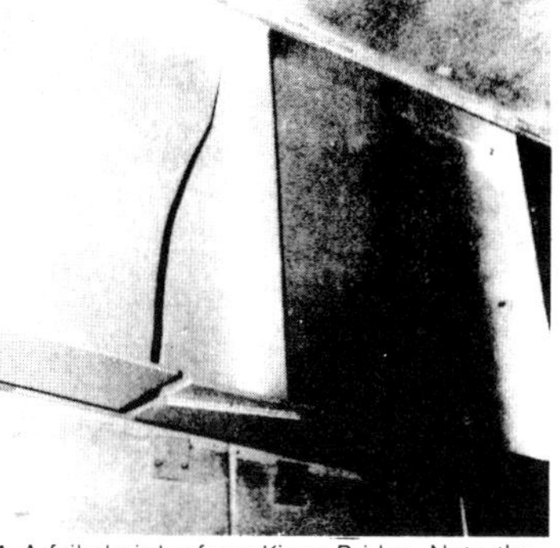

1 A failed girder from Kings Bridge. Note the fracture at the weld at the end of the cover plate

Mn increased the difficulty of producing crack-free welds. However, cracking was found to be so widespread that it was clear that the welding technique and subsequent inspection were also unsatisfactory. Inadequate preheat, possible use of wet electrodes and no predetermined welding sequence led to cracking which probably occurred either immediately after, or within a very short time, of welding. The evidence showed that welding and painting were taking place with little time in between, hence inspections had to be very hurried and must have been so cursory as to miss what turned out to be major defects.

The importance of strict control of composition and welding procedure to obtain crack-free welds is obvious to the competent metallurgist and these errors would not have been made had an able man been continuously connected with the project. A metallurgist, who was adequately knowledgeable on the problems involved, was employed as a consultant but, through no fault of his, contact with the specification authorities and with the contractors appears to have been spasmodic and inadequate for his knowledge, and concern about brittle fracture, to be transferred to the point where decisions were made both in inspection and construction. We find that, on the advice of the metallurgist, impact values for two temperatures were to be included in the specification. The specified impact values were probably not quite stringent enough; however, it turns out that the testing procedure was so unsatisfactory that the actual specification was almost irrelevant. The procedure often followed was to continue impact testing until the desired values were obtained! Subsequent tests on the steel from the girders gave low impact values at 0°C in the range 16 to 50 Nm Izod compared with the specified value of 27 Nm Izod. The failure occurred after a cold night (2°C) and at the time the lorry crossed the bridge it would seem that the steel plates were still in a temperature range at which they exhibited low impact values.

The major lesson to be learned is that the choice of materials and the advice on which that choice is made is of the greatest importance in making design and construction decisions and should not be left on an informal, even casual, basis which seems to have been the case on this bridge project.

Oil rig

In the case of our second example, the collapse of the drilling rig *Sea Gem*, a warning of the impending disaster was given. A few weeks before the collapse some of the tie-bars on which the rig was suspended from its legs broke. These tie bars were replaced, but it appears that no further action was taken to determine the cause of the failures.

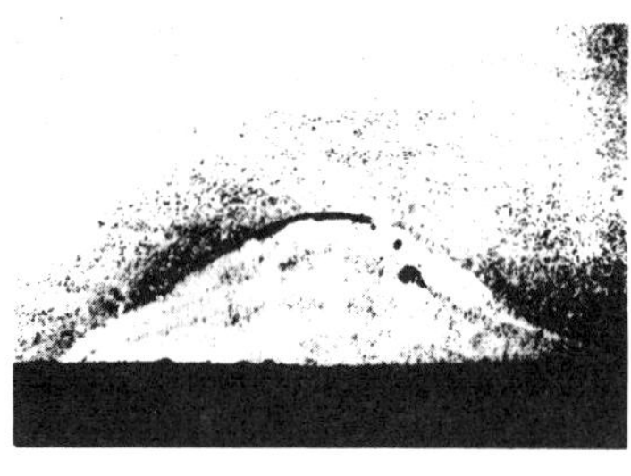

2 Macrograph of cracks associated with the repair welds on the tiebars

*The authors are with the Metallurgy & Materials Science Department, Imperial College, London.

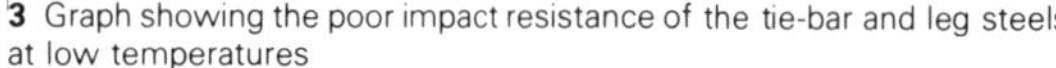

3 Graph showing the poor impact resistance of the tie-bar and leg steels at low temperatures

4 Failed pressure vessel

The tie-bars were gas cut from 7.6 cm steel plate; this led to some deep gouging which was made smooth by repair welding. There was a lack of knowledge of the dangers of low-input welds used for repair and some of these repair welds themselves developed shrinkage cracks which were obviously much more dangerous than the rounded gouge which they were supposed to improve (Fig. **2**). After the collapse detailed work on the steel of the tie-bar showed that at the temperature in the North Sea the impact values were much lower than would have been recommended at the time of construction (Fig. **3**). It is therefore not surprising that the stresses that would occur during the raising or lowering of the rig on its legs led to brittle fracture of the tie bars. These failures resulted in other tie bars being overstressed and failing so that failure spread along the side of the rig like a zip fastener being opened. The process of collapse of the tie bars led to bending stresses being applied to the legs which had both longitudinal and circumferential welds. All the legs were supposed to have passed radiographic examination but these welds contained major imperfections, such as lack of penetration and inadequate positioning of backing bars. Consequently brittle fracture developed in the legs, which were also constructed from a steel with low impact values at low temperatures (Fig. **3**).

It should be noted that this rig had previously been in service in the Gulf of Mexico and the Middle East before being adapted for use in the North Sea. Its early life, therefore, was in temperature conditions which would be favourable for the particular steels that were used. It was probably imprudent to use steels with such poor impact properties in any offshore environment; but, except for use in the North Sea, there was a relatively low risk of failure. On transfer to service in the North Sea there should have been sufficient technical awareness to recognise the very different conditions of stress and temperature. To this technical ignorance was added the ignorance of the likely behaviour of repair welds and the consequences of sloppy inspection procedures.

Pressure vessel

A few days before *Sea Gem* collapsed in the North Sea a pressure vessel being tested in the Midlands fractured catastrophically (Fig. **4**) and one segment weighing 2 tons was thrown a distance of over 45m.

Careful investigation revealed that the failure had been initiated in the region of the weld joining the forged end fitting and the rolled plate vessel (Fig. **5**). The origins of the two major fractures were microcracks in the heat-affected zone (h.a.z.) of the forging, which also contained a microcrack which had not propagated. The formation of these microcracks during welding was due to segregation in the forging. The cast analysis of the forging was out of specification as far as carbon and chromium were concerned, but this was known and accepted before fabrication of the vessel (Table 2). What was not known at that time was that the forging was heavily banded and that considerable segregation of carbon and manganese occurred within the bands. As a result of the high solute content, the hardness within the bands in the h.a.z. reached 400 to 470 HV in places. As the fabricators were unaware of the segregation, the preheat was discontinued immediately on completion of welding as is normal practice for homogeneous material with carbon levels up to 0.2%. The consequence of this action in this particular case was that hydrogen could not disperse to sufficiently low levels to prevent cracking in the hard banded regions in the h.a.z.

	Composition of forging percentage weight			
	Mn	**Cr**	**Mo**	**C**
Specification	1.5	0.70	0.28	0.17
Cast analysis	1.48	0.83	0.29	0.20
In band	1.88 to 2.00	0.80 to 0.82	0.32 to 0.39	0.25*
Outside band	1.53 to 1.59	0.69 to 0.71	0.25 to 0.21	0.20*

*Estimated from hardness values.

Table 2

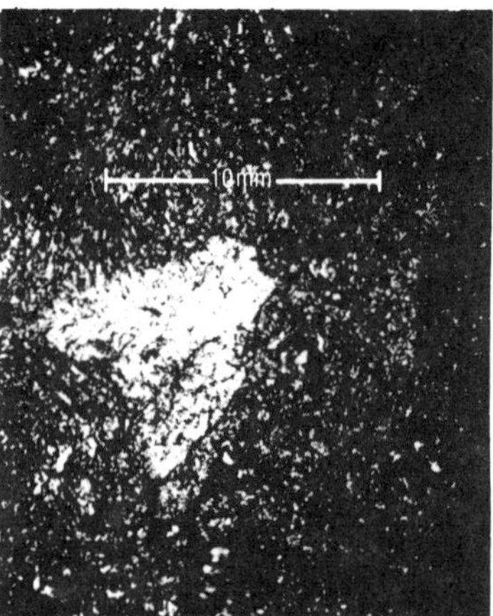

5 The origin of one of the main fractures. The bright facet lies in the heat affected zone of the forging and is thought to be initiation point

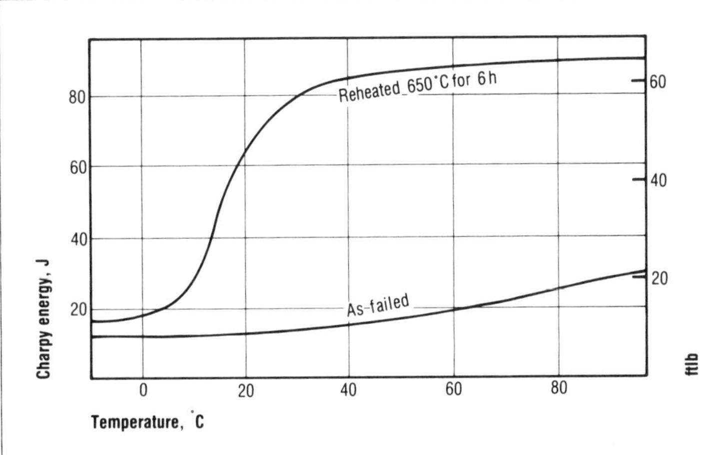

6 Graph showing the improved impact properties of the weld metal obtained by reheat treatment at 650°C for 2 hours

The formation of pre-existing cracks is thus accounted for, but why should they have propagated during testing? The initial stages of propagation were in the weld metal, which was found to have low impact properties. The vessel was supposed to have been stress relieved at 650°C to reduce residual stresses and to modify the microstructure of the weld. It was clearly demonstrated that the weld metal had never been at this temperature as reheating specimens taken from the failed vessel for six hours at 650°C markedly improved the impact properties (Fig. **6**). From experience gained since on stress-relieving similar vessels in the furnace it is estimated that the temperature of the weld in question only reached between 520 and 610°C dependent on position in furnace. Thus this failure clearly illustrates the importance of the control of segregation and heat-treatment temperature.

It is interesting to consider how the basis of technical decisions has changed in the intervening decade or so since these failures occurred. At the time of these failures it was not possible to make predictive calculations in terms of the actual stress situation and the size of the defects in a component. The advances in fracture mechanics have changed this situation so that considerations of safety are now on a quite different basis. Material parameters, such as the critical stress-intensity factor (K_{IC}), which indicate the susceptibility of a material to brittle failure, are now in use and are employed in a quantitative manner to determine critical crack lengths under given conditions of service.

There are a number of difficulties in applying fracture mechanics to a complex construction under service conditions. For example, the stresses in a complex construction with many attachments and openings is uncertain, the temperature of operation may fluctuate and the mechanical properties of the material, especially if welding has been employed, may vary. In spite of these problems the fracture-mechanics approach has been applied to reactor pressure vessels and it has been concluded (see the Marshall Report) that, within an acceptable risk, the pressure water reactor vessel would be safe. By inference, the same argument would lead to the same conclusion with regard to the steam drum of the s.g.h.w.r. Although, of course, practical experience of these very-thick-walled pressure vessels under nuclear conditions has not yet demonstrated whether these calculations lead to valid conclusions, the whole approach to the problem of fracture has changed since the major failures that have been referred to.

Hip prosthesis

The last case to be considered is that of fatigue failure. In the general sense, fatigue failures constitute by far the largest majority of minor breakdowns in engineering industry and will be familiar to most metallurgists and practising engineers. It is the failure of an unusual component which we wish to deal with, namely, a hip prosthesis used in making artificial hip joints.

The standard procedure is to hammer the prosthesis into the top of the femur and then to cement it in place with p.m.m.a. If the prosthesis is firmly fixed, the maximum bending stress should occur near to the base part of the prosthesis, which fits into a replacement socket in the pelvic bone. At this point of supposed maximum bending stress the prosthesis is of large cross-section and it would be intuitively supposed that the femur bone would break rather than the metal prosthesis under any applied stress.

Fig. **7** is a photograph of a prosthesis which has clearly fractured well away from the design point of maximum stress. Metallographic examination revealed that the fracture was transgranular and was not associated with any particular microstructural feature of the cast Co-Cr-Mo alloy. Fractography showed that there had been rubbing at one end of the fracture and that there were markings characteristic of a fatigue failure in this type of alloy. What clearly happened was that the fixing of the prosthesis in the femur was imperfect, thus allowing bending stresses to be applied along the shaft rather than in the thick region adjacent to the ball joint. As the shaft was tapered, as the point of application of bending moment moved towards the thin end the actual bending stress for a given body load increased. At a point about half way down the shaft this bending stress obviously reached a value at which a fatigue crack was initiated.

Other cases of fatigue failure of hip prosthesis are known and development of improved alloys for implants is continuing. The main priority of the implant material is, of course, that it should be compatible with the body. This is a severe restriction and only a few materials are suitable. Unfortunately, the commonly-used cast Co-Cr-Mo alloys have a poor fatigue resistance with a fatigue limit of only 200 to 300 MN/m². However, recent advances have been made in the fabrication of this type of alloy and now prostheses can be made from forged material with much improved fatigue properties ●

7 A failed prosthesis

Misuse of material

It can generally be stated that there are three reasons for material failure. These are: the wrong choice of material or process by the designer; the wrong treatment of the material by the producer and the misuse of the material by the user. There is little doubt that misuse of material in service is the most common reason for failure. Bob Ross* considers the problems most commonly identified in each of these three fields

Firstly, there is the designer with his blank piece of paper and the choice ranging from material such as titanium, with a tensile strength up to 75 ton/in^2 (1150N/mm^2), with corrosion resistance capable of withstanding the attack of hot concentrated acid, to mild steel with a tensile strength in the region of 20 ton/in^2 (200 N/mm^2), and very low resistance to corrosion, resulting in discolouration from a damp atmosphere and early failure by pitting and scaling in marine type atmospheres. The difference in cost between these two materials is a factor of 40 or 50 to 1, and thus the design engineer has to use considerable skill in choosing the correct material or treatment at the most economical price to cope with the conditions which the component will encounter in service.

Where low or medium strength is required and normal atmospheric conditions exist, then mild steel or low alloy steel will be the main choice. This can have a variety of treatments to improve the strength locally for specific purposes, and also to improve the corrosion resistance. The majority of engineering components make use of this type of material.

The design engineer, however, must be aware that properties such as impact strength will vary considerably within any one range of material and can in fact vary within a chemical analysis specification. Other properties such as tensile strength, ductility and corrosion resistance tend to vary less dramatically, but with corrosion there can be peculiar results with a very slight alteration in the corrosive environment.

Where high impact properties are required at low temperatures along with high or reasonably high tensile properties, then the choice of material becomes much more limited and the method by which the material is produced becomes of paramount importance. Similarly, when corrosion resistance to specific aggressive chemicals is required, then there is a range of materials capable of withstanding the specific corrosion environment, but the choice of material becomes much more limited with the necessity to have high strength along with corrosion resistance.

It is not the purpose of this article to discuss in depth the various materials which are available, but merely to point out the difficulties in front of the designer and the areas in which he is most likely to be mistaken.

Some sympathy should be given to the design engineer involved with low cost items. With this the economy of material and treatment become of vital importance, whereas with aircraft engineering, nuclear or other high cost items where safety is involved, then the design engineer has a better argument for using a superior material. It is also generally true that the detailed information available regarding the conditions under which the material will operate is more readily available in these areas than with consumer articles.

It is in the two areas of impact strength and corrosion that most failures resulting from design error occur. It is comparatively rare for materials of too low a strength or ductility to be chosen. It is, however, relatively common to find that materials fail in service because of impact and corrosion, where the design team were unaware of the peculiar problems which can exist with these parameters.

There is little doubt that corrosion failures are the most common where design is involved. There is considerable information to indicate that design engineers do not have sufficient knowledge of manufacturing techniques to choose the correct corrosion protection system, and also that they do not use the knowledge available regarding the application of these techniques. Very often components are produced with specific shapes or other parameters which make the correct application of a corrosion prevention system difficult, but by the time that this is discovered, the components have been manufactured, and the cost of re-design and manufacture must be considered before making any change.

Production abuse

The second cause of failure in the material field is where the correct material has been chosen and the correct processing or treatment specified, but this is abused during manufacture. There are four common treatments for metals: heat treatment, electro-plating, painting and welding. An inspection operation which will guarantee that the process has been correctly carried out is not possible after any of these treatments.

Contrast this with machining, where there are few exceptions to the fact that the method of production, whether it be milling, turning, grinding, hand filing or any other method of shaping metal mechanically will always result in physical methods of inspection being possible to decide whether or not the shape produced and the finish achieved is satisfactory.

Heat treatment for example, where steel is hardened and tempered, requires carefully controlled temperatures and rates of quenching. Unless these are complied with the result can be an increase in grain size which will seriously affect the impact properties of the material. This can only be identified with a destructive test. It is also possible to affect the proof or yield strength of steel and yet keep within the correct limits of its ultimate tensile strength. The only way of measuring the proof or yield strength of steel or any other material is by a destructive tensile test. The hardness test commonly used for inspection after heat treatment only measures the ultimate tensile strength.

From this it will be seen that to ensure that the correct heat treatment is carried out, it is essential to monitor the process throughout. Provided then that the correct material has been supplied there is little need for any final inspection.

With electro-plating it is possible to measure thickness locally but this does not ensure that the plating has good adhesion, that it does not have materials co-deposited with it which will affect the properties or that the deposit has been applied in a relatively homogeneous manner. To control these areas it would be necessary to do destructive testing on the finished component and even then there is no guarantee that the samples chosen would truly represent the batch being plated.

It is therefore essential in electro-plating to control the process sequence, to monitor the chemical analysis of the solution used, to identify the correct current density and to ensure that this is correctly applied. The time scale during plating is of paramout importance.

With painting there is no doubt that the preparation of the metal is of more importance than the application of the paint or the thickness achieved. As with electroplating, the measurement of thickness and adhesion tests will only supply limited information. Constant and careful monitoring of the metal preparation techniques whether it be blasting, wire brushing, etching, phosphating, etc, is essential and the time delay between this preparation and the application of the paint is of vital importance. There is no method of inspecting a finished painted component which will inform the inspector that a delay had occurred between metal preparation and application of the paint. This delay however, could result in paint failure after 6 or 9 months instead of an expected life of twenty years.

Welding can have expensive sophis-

*The author is Managing Director of the consultant metallurgists W H Herdsman Ltd of East Kilbride.

ticated inspection techniques such as X-ray, ultra-sonic testing, etc., carried out after completion of the join. These techniques will show faulty fusion, weld cracking or the presence of oxide, slag or flux. But there is no non-destructive technique which will show that the correct preheating has not been achieved and that a brittle layer exists in the heat affected zone. Only a destructive test involving impact tests, or a hardness survey across the prepared specimen, will identify this defect, which at one time was the most common welding problem.

It can be seen therefore that there are considerable areas in the control of metal processing where mistakes can and do occur. The method of overcoming these is firstly to prepare a detailed specification of the process involved. This is best prepared by the quality assurance department in conjunction with the production personnel involved. The document, however, should also be discussed with the design engineer who should be encouraged to communicate with production and quality assurance regarding any new processes, metal or design involving any of the above. This pre-production discussion can result in large financial savings, and also in the improvement of the process in the manipulative areas where the shape is difficult to heat treat, is impossible to plate, has surfaces which could not be readily painted or where a welder would not achieve correct penetration because of interference from other components.

Misuse by user

When the material enters service in the areas of aircraft engineering, nuclear and certain other limited applications, there will remain considerable control over the method and procedures in service. With aircraft engineering for instance, the control is documented. An example would be that an aircraft at take off has the power used stated for specific times and this is monitored and recorded. The same applies in nuclear engineering, and any deviation from the control procedures can be investigated by trained personnel. This means that failure of material in these industries is generally extremely limited.

With consumer goods the same control does not exist. High cost items will always have efforts made to have some control; for example, railway engines, heavy goods vehicles and marine equipment are subjected to routine maintenance. In some areas this routine maintenance is a special requirement and in others is a requirement prior to any meaningful insurance being available. These controls were found necessary in order to prevent expensive unscheduled break-downs, some of which would be dangerous to operators. However, they do not always result in preventing misuse, and other controls are now being examined, and in many cases used, to highlight problems before they arise.

One such technique is to examine at regular intervals the oil involved. This would be the lubricating oil from the engine, gearbox, etc. The sampling technique must ensure that a representative sample is taken and this is then chemically analysed for specific metals. With a knowledge of the metals involved in the engines, this analysis if carried out at regular intervals can highlight that wear is occurring in a certain area. This information can then be used either to stop the engine immediately and carry out an overhaul, or for detailed monitoring to ascertain if the situation is deteriorating. It must be emphasised that this is a diagnostic technique requiring routine examination and is of limited use as a one-off method of investigation.

The new technique (ferrography – see ENGINEERING, August 1977, p646) is now being commonly applied to aircraft, railway and marine engines and many engineering applications involved in civil engineering.

As the cost of the article is reduced, so the control over the use decreases. The normal motor car is probably the highest cost item used by the general public. When new, there is some control over the engine speed for a limited mileage. There is no method however, by which it is recorded whether or not these controls are exceeded, and provided the excess is not substantial no immediate defect will be seen but abuse could result in problems in early middle life of the car. Once out of this 'running in' period there is no control by the producer over the driver.

With the lower cost items such as hairdryers, hand tools and suchlike articles, then the control over the use becomes negligible. There are those of us who would always work within the instructions, making sure for example that an electric drill is not overloaded by using blunt equipment, is never stalled by overloading, has regular services and is kept clean. Others will never clean the equipment, will use blunt drills, will drop and gererally misuse the article.

Failure investigations under these conditions become of limited value as a detailed investigation will almost always identify some parameter which is slightly less than perfect. It is seldom possible to obtain detailed specification information and thus the investigator is left with the decision as to whether some slight deviation would be within the tolerance limits or whether it would be a serious defect. A common technique applied under these conditions is to compare the unsatisfactory or broken component with a similar component which has given a satisfactory life. It is very often only because the component being investigated has given vastly different results from a previous or similar component that any request for investigation is instigated.

Summary

To summarise then, failure of material is not often caused by design mistakes. It must however, be noted that, generally, when a design error is involved, it will result in a more disastrous failure than otherwise, or the failure will involve a large number of components.

In the manufacturing area by far the most common reason for failure is lack of control of the processes described in this article. The designer has a responsibility to ensure that any control necessary is specified. He cannot assume that 'Production will know'. In this area would also come the use of wrong material because of mistakes in the purchasing or storage of materials. This at one time was a common reason for failure, but in gereral there is now stricter control of purchase and storage of materials.

In the areas of processing until recently welding gave the greatest cause for concern. Welding on the whole was of an apparently low standard and though failures were relatively common, they were not consistent. With the advent of North Sea gas and oil however, welding failures of a dramatic nature have occurred causing some loss of life and considerable financial loss. This has resulted in a different approach to welding from the design stage, through the quality assurance, to the user himself. This has achieved a dramatic increase in the standard of welding throughout engineering in Britain. This increase in welding standards, in my opinion, is in the order of two to five times better than in the recent past.

The same cannot be said for corrosion control, this being mainly in the field of painting but also in electro-plating. Considerable advances remain to be made without in any way stretching our present knowledge or even improving on existing specifications. The need here is for engineering companies to insist on reasonable specifications being produced and then to ensure that these specifications are followed in the spirit if not the letter. Vast sums of money can be saved in this area. Heat treatment in many instances is correctly controlled but there remain considerable advances to be made in economical quality control, particularly on subcontract heat treatment.

Where the largest number of failures occur, that is where material in service is misused, then considerable education of the normal everyday person is necessary. This area is probably the most difficult and there is no doubt that many design engineers are thinking how improvements can be made to reduce the onus on the user. Whether this can be done in an economical manner remains to be seen ●

Corrosion behaviour of Ni-Resist cast irons

Owing to their relatively high alloy content, Ni-Resist austenitic irons exhibit a higher degree of corrosion/erosion resistance than ordinary cast irons or cast carbon steels in all environments. The influence of the nickel alloying addition increases the 'nobility' of Ni-Resist, placing it in the galvanic table between the nickel-free ferrous metals and copper-base alloys such as gunmetal and aluminium bronze. Ni-Resist irons vary in alloy content, to meet specific requirements, and are produced with both flake-graphite and spheroid-graphite structures. Practical experience indicates that in most applications the corrosion resistance of an SG Ni-Resist iron is similar or superior to the corresponding flake-graphite grade.

One of the main advantages of Ni-Resist irons is their good resistance to graphitization compared to ordinary grey cast iron. Although a thin graphite layer may form on the austenitic iron during service, the galvanic potential between the matrix and the graphite is less and any subsequent attack is reduced.

Quiet seawater

Table 1 shows the results of some corrosion tests performed in tropical seawater over a period of sixteen years. The corrosion rate for Ni-Resist type 2 is about five times lower than that obtained from grey iron and steel. It is accepted, also, that in seawater applications where the water is colder, as in the North Sea and Atlantic Ocean, the overall corrosion rate associated with all these materials is about 25 per cent lower than that obtained in warm tropical water.

Figure 1 shows a comparison between cast iron, Ni-Resist and 88/10/2 copper-base alloy in aerated, relatively slow-moving seawater. While the bronze and Ni-Resist exhibited similar corrosion resistance, the ordinary cast iron was very inferior.

Table 1 Long-term corrosion tests in seawater

(Data of C. R. Southwell and A. L. Alexander)

Duration: 16 years Location: Panama Canal	mm/a
Ni-Resist type 2	0·025
Grey cast iron	0·15
Cast steel	0·08

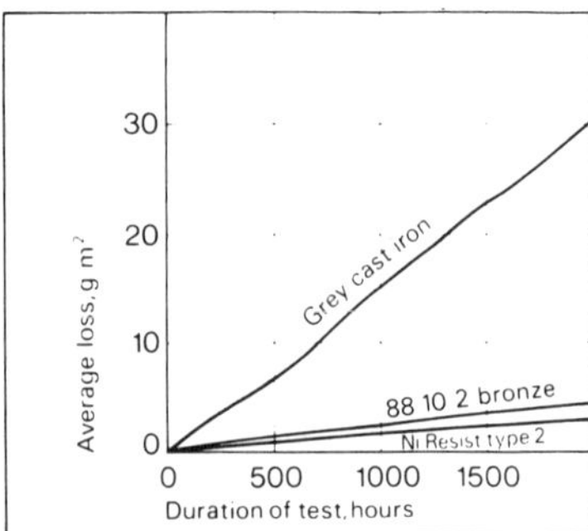

Figure 1. *Relative corrosion rates of Ni-Resist type 2, grey cast iron and 88/10/2 bronze, in aerated seawater.*

Table 2 Typical pitting rates in quiet seawater at 0·5 m/s

Material	Typical rate of penetration mm/a	Comments
Ni-Resist iron	0·05–0·13	slight graphitization
Aluminium bronze	0·08–0·25	dealuminifies
Grey cast iron	0·08–0·25	graphitizes

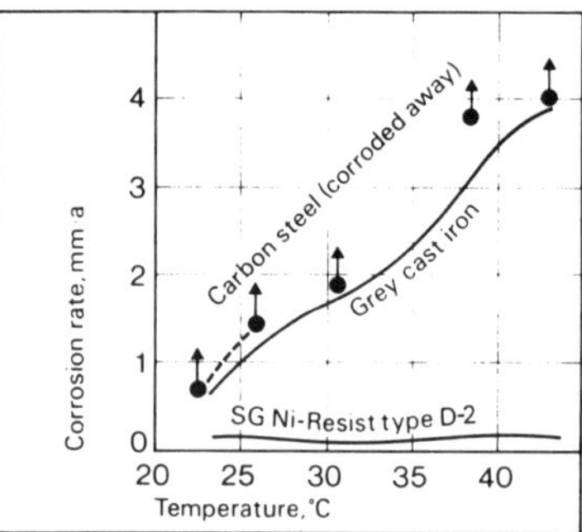

Fig. 2. *Corrosion rates of SG Ni-Resist type D-2, carbon steel and grey cast iron in aerated seawater – variation with temperature: continuous test exposure for 156 days.*

Table 3 E.E.S. seawater erosion test

Tests on bars rotating in seawater for 60 days with a tip velocity of 8·2 m/s. Seawater temperature was 28°C.

Material	Average corrosion rate per year mm
Ni-Resist type 2	0·75
88/10/2 bronze	1·0
85/5/5/5 gunmetal	1·8
Grey cast iron	6·9

Pitting

In quiet seawater, the rate of pitting penetration is at least of equal significance to the corrosion rate. Table 2, which gives some data on pitting rates obtained from tests in seawater, illustrates that Ni-Resist irons, besides being more resistant to pitting than grey iron and copper-base alloys, are not as susceptible to selective attacks, such as graphitization, dealuminification and dezincification.

Effects of temperature and aeration

The effects of increasing temperature and degree of aeration are most important in a number of applications, particularly desalination plants and cooling-water service.

A number of tests have been carried out in aerated seawater, over the temperature range 20–40°C, to compare the resistance of Ni-Resist type D-2 to that of cast iron and cast carbon steel. Figure 2 indicates the low increase in corrosion rate of Ni-Resist type D-2 within this temperature range, while the other materials suffered severe attack. Since similar results were obtained against concentrated brine, Ni-Resist is becoming widely used for the feed-water and brine-recirculation pumps in desalination plants.

Effect of velocity

The E.E.S. test was designed to measure the corrosion/erosion behaviour of materials in seawater. This test, developed by the U.S. Navy Experimental Engineering Section, rotates bars of material to provide a tip velocity of 8·2 metres per second. The results shown in table 3 demonstrate that Ni-Resist is superior to both grey iron and some copper-base alloys in high-velocity seawater.

Other corrosive media

Ni-Resist irons are widely used in the petroleum, petrochemical and chemical industries to resist corrosion from dilute acids. They are also useful in handling hot brines, concentrated caustic solutions and many other corrosive chemicals. Information is available on the resistance of Ni-Resist irons in many such applications and will be provided on request.

Gasket materials

by R. A. Weeks, James Walker & Co. Ltd.

The term 'gasket' has many interpretations, but, for clarity's sake, should only be applied to that component that provides a fluid-tight seal between two static components. Within the scope of this article, this has been further restricted to the seal between opposed flanges used for connecting assemblies containing fluid. For this function, two broad categories are available:

Jointings – sheet materials from which gaskets may be cut

Gaskets – custom produced sealing components to suit the individual flange.

Fig. **1** represents the forces acting on a static seal in a typical flange assembly and it should be noted that the force to compress the gasket initially (compressive load) must be sufficient to:

- seat the gasket into any imperfections in the flange or gasket surfaces
- overcome the hydrostatic end-force exerted by the fluid pressure
- leave sufficient residual stress on the gasket to contain the fluid pressure.

From this theory, it will be seen that available bolt-loading must be considered, along with size, pressure, media and temperature, as a factor in gasket selection.

Gasket materials in common use

Although a seemingly bewildering array of gasket materials and types is available, they actually break down into fairly well defined groups with readily defined areas of application. In cases of doubt, a reputable supplier will carry most types and the advice of their technical department should be sought.

Paper jointings: Plain paper may be used for dust seals or spacers, but, for fluid sealing, treated vegetable fibres should be used. Proprietary types possess good tensile strength, may be used on light gauge flanges and are suitable for service with oils, petrol, hydrocarbons and most organic solvents up to 120°C.

Paper is also available in the form of vulcanised (red) fibre but the extreme hardness of this form limits its use as a jointing.

Rubber jointings: Modern rubber technology provides a range of polymers to suit most industrial media with an upper temperature limit of 200°C (for the fluorocarbon types). Rubber is not a compressible material and will spread under load, a factor that limits flat gaskets to lower bolt loadings and, hence, lower pressures of the order of about 3.5 bar. The addition of a reinforcing cloth or metal mesh between rubber plies will reduce spread and allow some pressure increase.

Alternatively, the flange configuration may be adapted to restrain the rubber and this is most effective in the form of a simple recess containing an 'O' ring which, in its simplest form, may be used up to 105 bar.

Bonded cork jointings: The use of cork in gasket materials is to provide a cellular structure which will compress within itself thus providing a jointing material suitable for low bolt loading and/or irregular flanges.

Protein or resin bonded cork is a low-cost material suitable for light duties (e.g. automotive rocker box or sump gaskets) but for increased strength and wider range of usage, a rubber bonded cork should be specified. The presence of a rubber binder ensures that a seal is readily established at very low seating stresses and, even under higher clamping loads, there is no tendency towards the excessive spreading associated with solid rubber.

Chemical compatibility is largely decided by the binder and most brands are suitable for oils, solvents, greases, refrigerants, water, air, etc., in the temperature range −30 to +120°C. High-temperature polymers can extend this to about 150°C, which represents the limit for the cork.

Compressed asbestos fibre jointings: Fibres of white (chrysotile) asbestos are opened out and thoroughly mixed with a rubber-based binder prior to sheeting out between opposed rollers to form the family of jointing materials collectively known as compressed asbestos fibre. The quality of the asbestos fibres and the proportion and nature of the binder mix will determine the characteristics of the finished jointing. Grades are available for most media up to a limit of 510°C for the premier types.

These jointings are fairly hard and should only normally be considered for machined flanges since a residual load of 15 to 30 MN/m² of gasket surface is generally required for a seal.

Within these limitations, this type of gasket is capable of efficient and economical sealing for all but the most arduous conditions. It should be noted that flange surfaces should be such that a gasket thick-

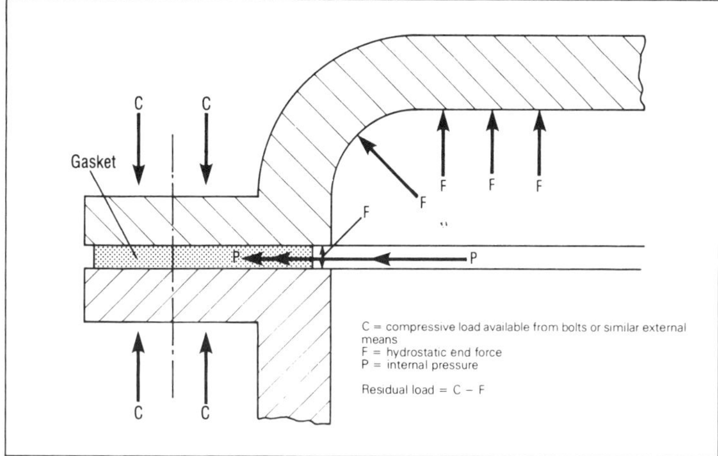

1 Static sealing

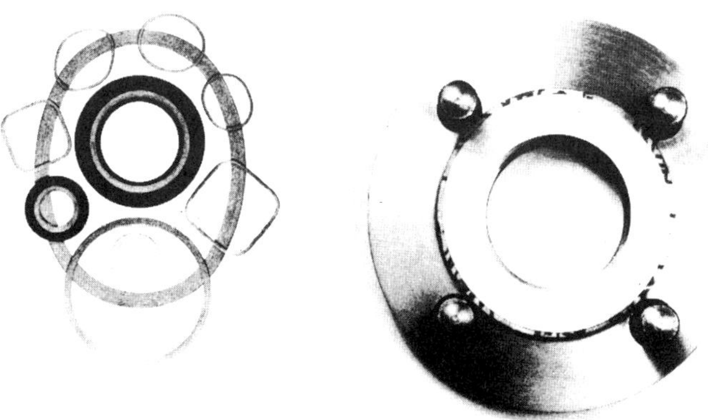

2 Spiral-wound metallic gaskets

3 P.T.F.E. envelope gasket

ness of 0.5 mm can be selected since thicker gaskets have a tendency to creep, which will limit their pressure capabilities.

Wire reinforcement may be added to the sheeting for increased creep strength (particularly for narrow flange widths) but this does not increase the jointing's temperature resistance.

Woven asbestos jointings: Rubber proofed asbestos cloth for lighter bolt loadings and temperatures up to 230°C (e.g. boiler manhole joints or fabricated ducting flanges). Also available is a woven twill of stainless-steel wire and asbestos with heat resistant binders which may be used on dry gas sealing up to 815°C.

Metallic gaskets: Flat 'washer' types require high compressive load and good surface finish but the use of shaped sections (round, diamond, delta, serrated or grooved, API ring-joint type and other variants) is intended to obtain sealing contact within practical bolting limits and gain the strength and security of a metal-to-metal joint.

Metallic sheaths may also be used to give protection to more resilient jointing types, e.g. copper covered asbestos cylinder head gaskets.

Spiral-wound metallic gaskets: Produced by winding metallic strip on edge with an interwinding of filler material (Fig. **2**). The windings are usually deformed into a horizontal 'V' configuration. While requiring fairly high bolt loading, these gaskets possess the important feature of resiliency thus maintaining a seal under conditions of thermal cycling or vibration. Solid metal rings on either side of the windings are preferred for protection of the sealing element, location against the flange bolts and as a means of preventing over-compression.

These gaskets can be readily used at temperatures from the cryogenic to at least 1000°C and pressures up to at least 350 bar. The metals and fillers may be varied to suit the application requirements.

P.T.F.E. gaskets: P.T.F.E. is unaffected by most chemicals and has a useful temperature range from −200 to +250°C: ideal properties for a gasket material. Unfortunately, it lacks resilience and is subject to cold flow under compressive load. A seal can be obtained with gaskets cut from p.t.f.e. sheet but the flange arrangement must ensure retention of at least the gasket o.d. (e.g. a spigot and recess).

However, the properties of p.t.f.e. can be used for gaskets in the following forms:

P.T.F.E. envelope gaskets (see Fig. **3**), where an outer sheath of p.t.f.e. is used to protect a gasket cut from traditional materials.

Extruded p.t.f.e. cord or tape braided from p.t.f.e. yarn can be used to form readily compressible gaskets on site.

P.T.F.E. can be used as the filler material for spiral wound gaskets ●

If you have a
etallurgical problem,
our wife may be
e first person you
hould talk to.

If you make products that involve metal, and if that metal fails under the conditions it is subjected to, the answer may be nearer than you think. Watch your wife using a stainless steel saucepan and let your mind go to work.

That pan has to cope with fierce heat and sudden temperature changes. It gets scraped and scratched by spoons, forks and knives. It gets a thorough scouring every time it's washed up.

And it stands up to the lot. That's why Prestige make their handsome saucepans from BSC stainless steel. Partly for its gleaming good looks...but mainly for its long life of absolute hygiene where food preparation is concerned. (Prestige also guarantee every piece for ten years!)

So if you are involved in designing with steel or aluminium, brass, or copper, think again about stainless.

Of course, it can cost more initially. And by increasing the materials content, you push up your price. But don't dismiss stainless until you've done your sums right through, because often you'll find two things.

The longer life of the product makes the added cost worthwhile.

And you gain the two extra selling points of higher quality and cheaper maintenance.

Yes, think again about stainless. Find out the current facts about our range of thirty different types. And remember, our back-up service is always at your service, particularly in matching the performance of our steels to your exact needs.

Write to BSC Stainless Marketing,
PO Box 150, Sheffield S9 1TQ.

The cost of corrosion The Hoar Report* estimates Britain's losses from corrosion as costing us a horrifying three-and-a-half thousand million pounds.

Much of this loss is preventable. Stainless steel is the supreme example of an existing material that must be used more fully for its superb resistance to corrosion.

And British Steel has already invested £130 million in plant to double our capacity to supply it.

*"A Survey of Corrosion and Protection in the UK," published by the D.T.I. in 1971 (figures adjusted for inflation).

The material you've been looking for could be right at your fingertips.

Corrosion war

Many ways have been developed to combat body corrosion in the motor car. M A Jacobson, chief engineer of the Automobile Association, discusses the effectiveness of some of them

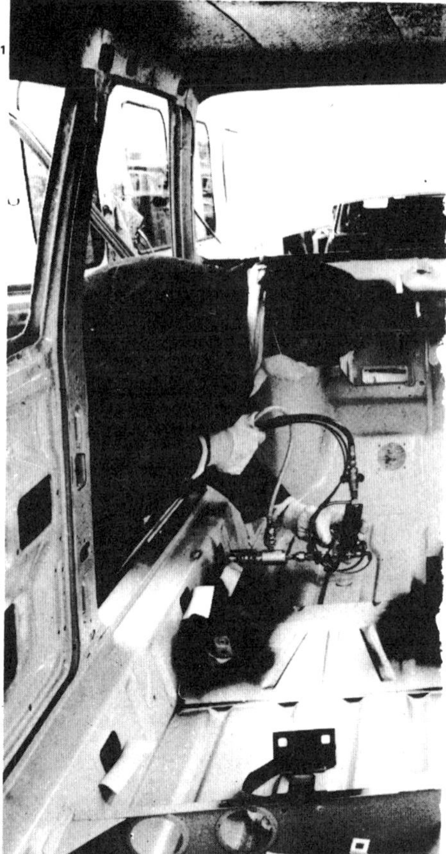
1

2

Better body design, improvements in metal pretreatment (particularly better phosphating control prior to primer painting), and electrophoretic primer paint deposition have helped to give an economic life of 10 to 11½ years to the all steel car body in to-day's very corrosively aggressive operating environment. Additional processes by manufacturers and specialist anti-rust treaters could give a potential extra life of between 2 to 5 years to the structure provided they are periodically checked and given a booster type of additional material deposition where required. Many of these processes have a basic service life shorter than the steel structure they are supposed to protect.

There are four different zones where special compounds are applied either on the car assembly line or subsequently by specialist firms. These are:

- the underbody, including the underwing areas, and possibly the lower part of the outside of body sills;
- internal surfaces of hollow sections;
- between trim and painted panels; and
- the engine compartment.

On the underbody some manufacturers use compounds with a significant proportion of short asbestos fibres or similar material which have no merit as far as rust prevention is concerned. These fibrous materials are effective sound deadeners but their main purpose is to protect the factory-applied paint against erosion by stone and grit bombardment.

External coatings

In rising order of durability externally applied coatings fall into one of four main groups:

1. Wax compounds (these may be long life protection types or give only 6-9 months temporary protection). Wax or petroleum based heavy compounds applied externally often are derivatives of compounds which are injected into hollow sections. Some also contain a highly refined fully processed (oxidised under controlled conditions) "blown" bitumen to give the compound some body. They are solvent based and take anything from ½ hour to about 3 to 4 days to set to their durable consistency, depending on their compositions and the temperature of applied compound and body shell. The materials are generally thixotropic. Build up at about ½ to 1½mm thickness is fairly common. The covering should never set hard – hence there should be no danger of age-embrittlement.

Many of the proprietary components have a good abrasion resistance and are partially self healing – that is they flow back into areas partly eroded by slush and grit. But their penetration into body seams and nooks and crannies is poor. Some have moisture displacing properties. Providing they are of the same family group, it is possible to key subsequent coats into older ones with success, provided dirt and grit and loosened or damaged sections of the compound have been previously removed by high pressure water hosing and that ample time has been allowed for drying before application. The solvent contained in the re-application partially softens the original compounds, and during air drying allows the two to blend and thereby form a covering impervious to water and oxygen, the two constituents essential for initiating rusting of sheet steel.

2. Bitumen compounds The cheapest compounds (and hence the most frequently found on many UK produced cars) are bituminous. Many of the cheap water-based ones give a disappointingly short anti-rust protection, for they are often significantly hygroscopic due to partial curing if they have been applied in quite thick layers; as a result they allow corrosively aggressive moisture to be retained in cracks and crannies.

Bituminous coatings, even the better solvent-based ones, offer very poor resistance to fuel leaks and are softened and removed by oil and grease spattering. Most of the factory-applied bituminous compounds eventually age-harden and crack due to oxidation. Many car manufacturers apply them too thickly before the final stoving of the painted body shell. The rapid application of heat causes them to flow and then to shrink often into quite a bizarre surface texture with a brittle outer skin. Less compound would be better. The reasons for over application are that it is a messy, unpleasant, and difficult-to-control process, secondly, more compound adds to a quieter ride.

When age cracking and partial separation from the sheet occur (usually after 3-5 years) all loose, cracked and improperly keyed material must be removed completely and major sections will then need re-treatment.

3. Rubber compounds The rubber-based compounds are also black and usually are good as sound deadeners, even without the addition of short, fibrous material. They can be factory applied or as a post treatment and generally retain some flexibility.

But butyl rubber or similar material is more difficult to bond effectively on to painted steel than the bituminous compounds. The surfaces have to be much cleaner. When properly applied, the coatings do not age-crack or shrink, have good erosion resistance and give excellent anti-corrosive protection. But they are a significantly more expensive product than bitumen based ones. Service life is about 5 to 8 years, ie longer than bituminous compounds yet less than a good PVC-type plastics coating. But it is possible to apply a supplementary coating over the original compound, thereby extending overall service life. Several vehicle manufacturers compromise by having a rubberised bitumen.

4. Plastics coatings Thin plastics coatings which, despite their higher cost and lower sound deadening properties, have been popular with many European car manufac-

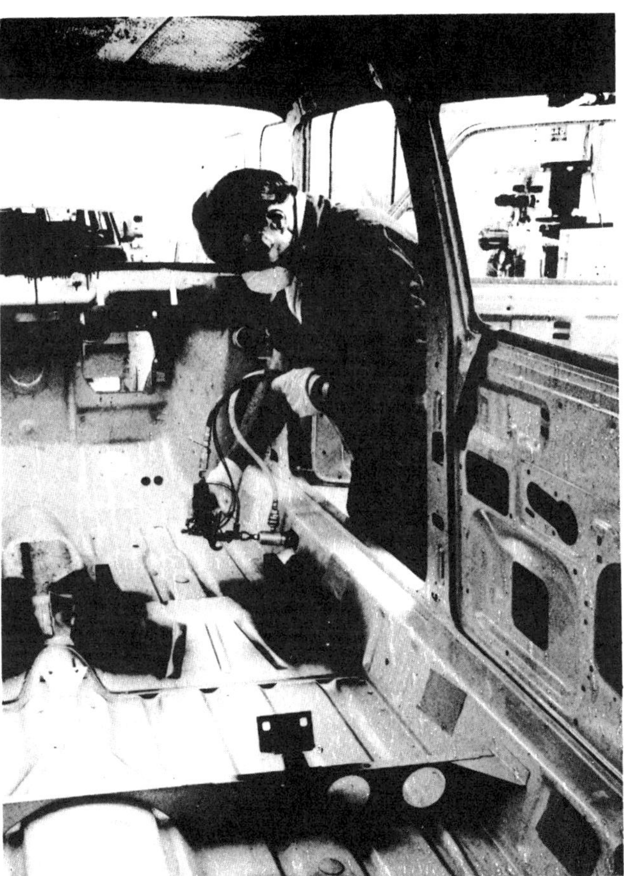

1 Wax injection being carried out at British Leyland. **2** Protectol's non-drip Gold Seal, being applied to a Dolomite wheel arch. **3** Flaking exposes the bare metal. **4** Underbody sealant is fan sprayed by Protectol

4

3

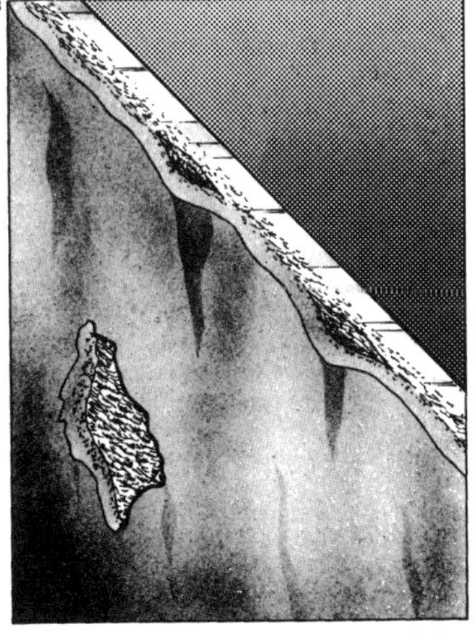

turers for a number of years, are gaining ground, because such a coating is aesthetically more acceptable. It will readily take an overlay of most automatic top coat paints and offers a longer life of corrosion and erosion resistance. It is generally a very thin coating, hence, unlike the heavier types, expands and contracts in complete unison with the steel body shell under varying service load conditions.

However it cannot easily be repaired, since it is cured by fairly high stoving temperatures during the painting of the body. In the case of PVC it needs to be above about 115°C to prevent the formation of monomers and hydrochloric acid at the free boundaries. Such temperatures are easily achieved on the production line but cannot be reached in after-service repair trade facilities – even where low bake ovens are in use. Strict regulations must be observed by the manufacturers to avoid a health risk to the operators.

Internal coatings

Not many years ago some vehicle manufacturers considered it barely necessary to protect, by spray phosphating, the inside of hollow sections. Since then phosphating followed by primer coat application, achieved by dipping the body in a bath, has become the standard painting process. More and more car builders now spray-in petroleum or wax based compounds under pressure, using special long lances ending in nozzles, which allow for a thin, even coating of the inside of panels.

Spray pattern of the nozzles, pressure setting, air assisted gun or airless application, solvent content, viscosity and chemical composition vary from one proprietary process to another, but they all have a number of common factors.

A relatively low flash point.

A high solvent content.

They are air drying in a moderately temperate environment by solvent evaporation. but never set hard. If too much compound is applied in the first instance or in a subsequent treatment, then the nuisance of dripping from joints or drain holes, may continue for a long time, particularly in very hot weather.

They penetrate into close fitting joints, nooks and crannies – even against gravity.

They displace a moderate amount of surface condensation and after becoming tacky prevent moisture and air getting at the metal to which they adhere.

It is this last property which gives them their anti-corrosive merit.

But here is an inherent problem. Such thin protective films will not adhere to a surface covered by loose dirt and dust, oil, paraffin, or silicones, for they do not react chemically with paint or metal but purely by the active polar attraction agents contained in their formulation.

It is therefore pointless to apply them to anything already suffering from substantial surface rusting, particularly if it has reached the advanced stage of flake rusting, unless the rust is first effectively removed mechanically or chemically. Due to their polar agents the films will key into the irregular rust flakes in preference to the parent steel, and can then lead to a localised but more rapid corrosion of the parent metal where such covering is non-existent.

To prevent bimetallic corrosion and mechanical fretting, an application of such compounds, of fairly high viscosity, between trim and painted panels can be beneficial.

The engine compartment rarely requires anti-rust treatment, except for the body seams and overlapping joints. However, many of the specialist firms apply a dark brown or black coating all over, for appearance's sake and since it gives some limited additional sound damping.

Anodic protection

By making a protective layer anodic (negatively charged relative to the steel structure) it is possible to ensure it is only the protective layer which dissolves slowly in the presence of slush, salt, mud, etc. and thereby prevents rusting of the sheet steel underneath it. Pure zinc and aluminium are such sacrificial metals. One process which, though somewhat expensive, has been adopted by some Continental specialist rust treaters, involves 'silvering' the car. In fact it does not contain silver at all, but it is a covering of practically pure aluminium to

Corrosion war

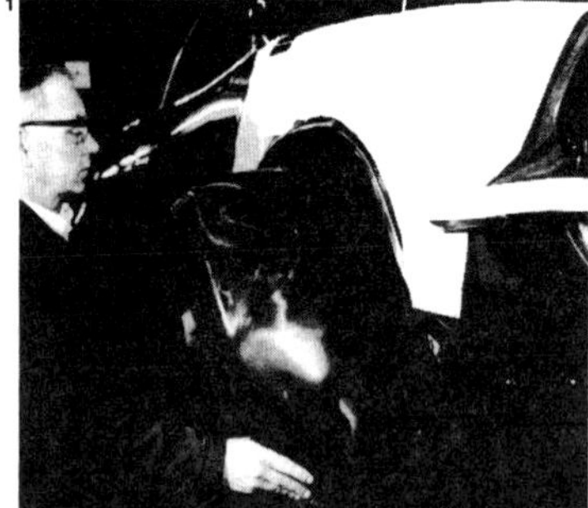

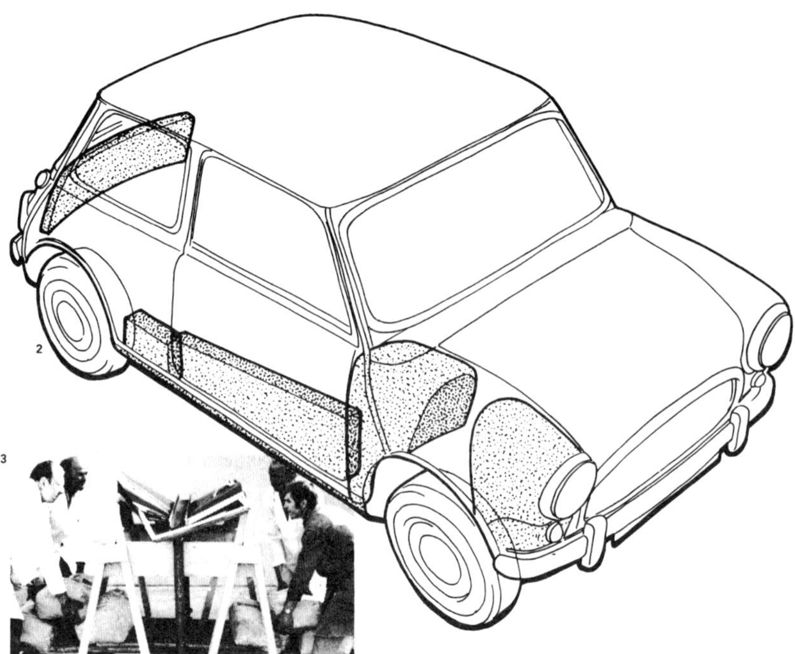

1 Polypropylene wheel arch liners being fitted to a Princess. **2,3** Internal sections filled with Bondafoam give both protection and increased strength.

the underside, wheel arches (engine and boot compartments if required) and also hollow sections. Being capable of withstanding much higher temperatures than the conventional bitumen or wax and petroleum based materials, there is no need to mask exhaust silencers and pipes. The 'silvering' also reduces accidental fire risk during subsequent cutting and welding operations.

The material provides a thin sacrificial metal coating all over. When corrosion occurs, the aluminium rather than the ferrous metal components are attacked. Further application of the sprayed-on aluminium, after the original deposit is eaten away, restores the protective power – which is good. But it is an expensive process.

The Vauxhall system of injecting a silver-coloured bituminous aluminium compound into the body sill section does not use a truly anodic sacrificial compound. It does however, seal the interior of the box sections against water and air, and permits easier on-the-line inspection.

Sacrificial zinc out of an aerosol can be applied on to bared metal by the DIY enthusiast – and provided it is almost 100% pure metallic zinc and applied carefully can be very effective, particularly if it is then followed by a primer paint to seal it.

As with all new processes, some do pose problems to the garage trade during body repairs. Repairers should be aware that they may have some fume and fire risks during cutting and welding due to partial internal wax injection or similar treatment of hollow sections and flammability of the underbody protection coating or foam fillings in some sections.

Effects of body repairs

A good deal of the anti-rust protection is all too often lost when major body restoration is undertaken after an accident. Provided the rest of the metal structure is reasonably free of rust, it is possible to carry out successful welding-in of new panels, even of major structural sections. Separate panels and reinforcing pressings often arrive at the repairers with factory-applied phosphate and primer coat, but during the cutting and subsequent welding operations anti-corrosive protective layers are often charred or locally destroyed. Most repairers do not bother to counteract such damage by some special edge primer treatment, for it often does not manifest itself for up to a year or two after the repair has been passed as sound by an Insurance Assessor.

Most body repairs lead to early onset of rust, for a charred protective layer means a broken bond of phospate and primer paint to the sheet steel, and, while it is possible to restore a good deal of the anti-rust protection on the outside of panels by a proprietary type of etch primer, this is impracticable on the inside of hollow sections. Where sections are spot or seam welded, the primer coating will be destroyed locally, leaving tide-line demarcation for subsequent preferential attack by corrosion.

In our aggressive corrosive environment there will always be enough concentration of some electrolyte and electrical potential to form and sustain localised corrosion cells. Dust and moisture, containing some industrial fall-out or road-salt, are bound to penetrate even into fairly close fitting joints. The electrical potential and the electrolyte covering may be insufficient to sustain corrosion over a large area but quite effective in a small concentrated area of bared metal. There are many flanged joints, for instance on wing sections, where, prior to spot welding-in the outer panels, it is feasible to apply weld-through plastisol-type corrosion retarding sealer strips both on the assembly line and in the repair shop.

Matching protection after accident repair with the original factory processes can be a problem. For instance PVC on certain sections (sills), bitumen on others (the main underfloor and underwing areas) are all stoved-in on the line. It is impracticable to re-apply PVC and high solvent bitumens, and high-solvent wax or petroleum based underbody compounds permanently soften the remaining PVC by plasticiser migration. The solvent may also loosen the keying of the PVC at its edges, and it can then be peeled off. A heavy deposit may even cause the large sections to drop off. The vehicle manufacturer's approval of complete compatibility over an extended life should be obtained before spraying or brushing a particular brand of proprietary coating on to a factory-applied one.

Foam filling of difficult-to-protect body sections is likely to be on the increase. The very lightweight closed cell foams – usually polyurethane – are meant to fill such hollow sections completely. They stiffen them, reduce noise, vibration and slight fretting movements and absorb impact energy in collisions. However if the process results in only partial filling or the foam lacks proper adhesion to the metal, it can easily become one of the quickest ways of causing inside-out corrosion.

The motor industry usually finds a way round its problems. Good examples are underwing protection methods. The fitting of complete inner wheel arches having aerodynamically designed free air flow on both sides of the cheap smooth plastics liner has eliminated packing by mud and slush. If the ventilation of the back of the liner is good, the net result is excellent. If insufficient thought and no airflow studies have been applied, then this concept – aimed at eliminating erosion and corrosion resisting underwing treatments – can have most disappointing side effects ●

Steel-reinforced concrete

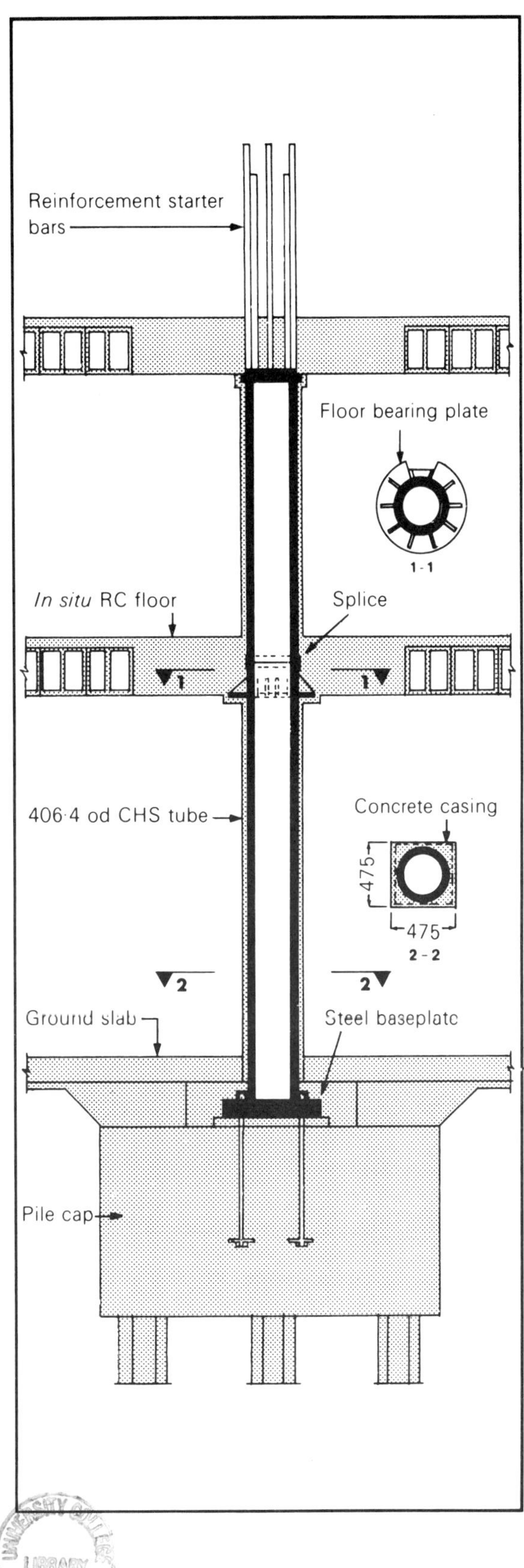

Introduction

The shortage of land suitable for building development, together with an associated rise in its cost, has produced the economic necessity of constructing taller and taller buildings. This in turn has stimulated the development of specialised erection techniques.

Design Constraints

The constraints imposed upon structural systems for multi-storey buildings fall into three categories:–

Requirements for Structural Safety and Serviceability

For example, the structural framing must be sufficiently strong to transfer the vertical and horizontal loads to the foundations and be sufficiently rigid to limit the vertical deflections and horizontal sway. Besides containing horizontal and vertical movements (perhaps brought about by differences in temperature and wind loading) the structural framing must also be durable and capable of providing adequate fire resistance.

In a multi-storey building circulation of people is predominantly vertical, so the provision of lifts and staircases warrants particular attention. The need for air conditioning and other essential services makes demands on space within the structure.

Planning Considerations

The type of occupancy for which a building is being planned, for example, will govern the magnitude of the floor loading and strongly influence the geometrical configuration of the frame. This aspect too will have considerable bearing on the degree of internal planned flexibility required.

Frequently a multi-storey building will be constructed on a confined site or one to which access may be severely limited. Considerations of this nature will impose limits on the degree of pre-fabrication possible and the size of component which can be transported to site.

Economic Considerations

Questions of structural feasibility are followed by those of economic viability. Building costs comprise two elements; capital costs of construction and recurring costs arising from the operation and maintenance of the building. Besides the cost of the frame itself, the choice of a particular framing system will influence the individual capital costs of foundations, cladding, HVAC installations and fire protection. Furthermore the speed with which the frame itself can be constructed will bear upon the cost of construction.

The costs of owning and operating a multi-storey building are dictated partly by the decisions of its designer. 'SHS the Builder' illustrates ways in which structural hollow sections have helped designers to meet the requirements mentioned above.

Glasgow Royal Infirmary Redevelopment

Phase I of the eleven storey Glasgow Royal Infirmary Redevelopment was designed for in situ reinforced concrete construction. Investigation showed, however, that some columns would be too highly loaded in reinforced concrete to meet the maximum desired column size of 500mm square. Thus 70 columns, from a total of 1700, required further investigation.

It was found that 406·4mm od CHS in Grade 55C steel could carry the required design loads – leaving adequate room within the desired 500mm limits for a fire-protective casing. Design proceeded on this basis and a special rolling of 406·4mm x 34·9 and 38·1mm heavy wall CHS catered for the highest design loads of over 11000kN.

The lower ends of the columns terminated in thick steel base plates which spread the loads into the pilecaps through high strength grout (60N/mm^2 at 28 days). The CHS thicknesses varied with load, changing at storey heights where the joint was made by butting the machined ends of the sections together and enclosing with a collar plate welded to both sections. Such welding provided the option of erecting one storey at a time and welding in situ or prefabricating in the shop. In the event, the sections were welded together in the shop for erection in units of up to three storeys in height (12m) cantilevered from the pilecaps by the holding down bolts. The transition from CHS to reinforced concrete column construction was achieved by welding a prefabricated cage of reinforcement starter bars to the steel column cap plate (the only site welding involved). Bearing for the in situ reinforced concrete floor slabs was provided by annular steel plates welded to the perimeters of the tubes and stiffened with welded gussets.

The choice of fire-protective covering was dictated by the need to provide a hard casing, of the same external dimensions as the reinforced concrete columns, which would allow firm fixings to be made to locate and stabilise partitions. Concrete reinforced with a light mesh was used, giving a fire rating in excess of two hours.

Client:	The Greater Glasgow Health Board and the University of Glasgow
Architect:	Sir Basil Spence, Glover and Ferguson in association with TDW Astorga, Central Services Agency, Scottish Health Service
Structural Consultant:	Ove Arup and Partners
Main Contractor:	John Laing Construction Limited
Tube Column Subcontractor:	Pipework Engineering Developments, Tubes Division, British Steel Corporation

Ceramics in turbine design

Early in the nineteen seventies the Advanced Research Projects Agency in the United States launched an adventurous programme of research and development. Their aim was to demonstrate that brittle materials, that is ceramics, would be successful as components in high temperature gas turbines. Specifically they chose to demonstrate this on a small scale vehicular turbine and on a large turbine used for power generation. Ford were to be responsible for the small engine whilst Westinghouse were to investigate the large component design. This covers both ends of the power spectrum, from 300 hp in the vehicular turbine to 40 000 hp in the generating turbine, to demonstrate the potential for engines of any size within this range. The announcement of this project combined with its giga-dollar investment caused consternation amongst turbine engineers.

A similar programme had been carried out in the early 'fifties using cermets* as potential blade material. This work showed the disastrous consequences of using brittle materials in highly stressed moving parts. Because of the material's low toughness any small fragmentation during failure of a blade impacted the remaining blades and caused the whole rotor to fail.

Later investigations have shown that the project failed mainly because the blades were made to almost a Chinese copy of a metal blade and as such were endowed with many stress concentrations which are not so critical in a ductile super alloy. This and other reasons revived the interest in ceramics for turbines and so the ARPA programme goes ahead – but why bother with ceramics in the first place?

* These materials comprise a hard ceramic phase such as tungsten carbide bonded by a metal matrix. These are the materials used as machine tool cutting edges.

This is the third in the series of articles linked with the Open University course 'Materials Under Stress'. The associated programme entitled **The turbine blade 3** will be broadcast on BBC2 at 1215 on 9th October and repeated on BBC1 at 0730 on 12th October.

The programme is the third of a series within the course dealing with turbine blade design. The first two programmes will be broadcast at the following times:
The turbine blade 1 – Different blade, different alloy, different process BBC2 1240 11th September and BBC1 0730 14th September
The turbine blade 2 – Predicting failure BBC2 1240 25th September and BBC1 0730 28th September.
All three programmes will be presented by Ron Jones*, author of this article.

*Lecturer in Materials Science, The Open University.

Advantages in using ceramics

The major advantage in using ceramics is that the components can operate at much higher temperatures than super alloy parts. In fact the ARPA programme is aimed at demonstrating their use at 1370°C. The consequences of running at such high turbine inlet temperatures are an increase in turbine efficiency and a reduction in specific fuel consumption. Also because these components would operate at this temperature uncooled there would be a power increase; in the case of the small engine this would mean doubling its present capabilities.

Other reasons are that the engine could run on a low grade fuel. At high temperatures, low grade fuel can be burnt efficiently. Low grade fuels, however, release impurities which often cause rapid attack and corrosion in metal components. Ceramics at elevated temperatures are far more inert to these corrosive environments.

Finally, the raw materials on which many engineering ceramics are based are far more abundant than the metals used in super alloys, so the usage of these materials does not represent such a threat to world resources. Ultimately this has an economic effect in that it determines the raw materials cost; Fig 1 shows a comparison of raw materials cost between six engineering ceramics and the super alloys. Such advantages provide a strong motivation for a programme of work such as this. This article gives an indication of the systematic approach to the design problems and the success of the project to date.

The small vehicular engine

Still singing the praises of ceramics, here are more facts and figures for the Ford vehicular turbine. There is an engine which could be used now. It is not a ceramic engine and employs conventional alloys. The rotor blades are cooled as are the stator vanes and the engine operates at about 1100°C. For a compression ratio in the range of 4-6:1 an increase in the operating temperature to 1370°C would give an improvement of 20% in specific fuel consumption. This, coupled with doubled power output, makes the small ceramic turbine a very desirable alternative to the existing internal combustion engines used in road transport applications. It is not sufficient, however, to simply replace existing parts in the Ford engine with ceramics. Some new and rather revolutionary design philosophy is required.

The disastrous consequences of blade failure in a ceramic turbine engine are perhaps the only undesirable quality of cermics. Unfortunately this brittleness is an

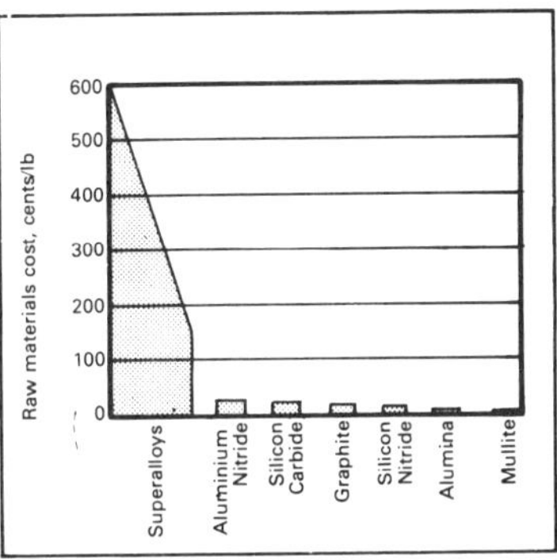

1 Raw material costs of high-temperature materials

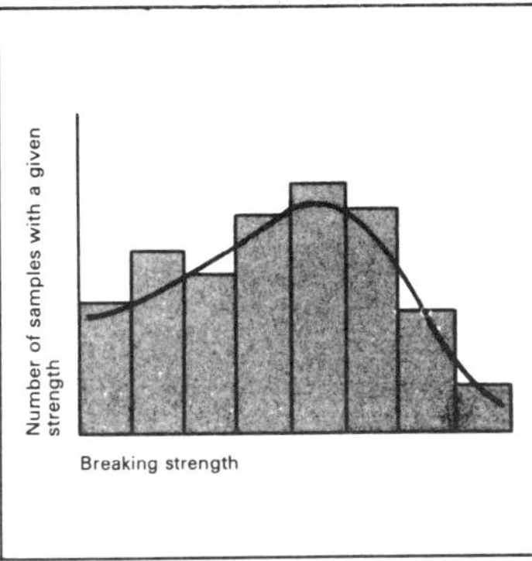

2 Bend test histogram

inherent property of ceramics and although materials scientists have made some progress towards developing ceramics with greater toughness, the values obtained are still many orders of magnitude below those of most super alloys. This looks like a dead end for further research so the ceramics designer has chosen the alternative path: "The ceramic must not be allowed to fail in service".

What exactly does this mean? The component must be designed to a safety factor sufficient to allow for no failure to occur over the period of service. This in turn means the component must be strong enough to withstand all the possible stresses suffered in operation. Whilst this may not sound too tall a task to those unfamiliar with ceramics, it is known to the initiated to be far from simple. This is because of the wide variation of strength occurring in any sample of a ceramic material. For instance most engineers are satisfied with a tensile strength value for a metal based on the average of three or four test results. Ceramics, however, can show two orders of magnitude difference in breaking stress over a handful of test pieces made from the same batch. This forces the designer's hand somewhat, and means that statistics must now be employed to describe the strength of these materials.

The reason for such scatter is that these materials are very susceptible to the presence of tiny flaws or cracks and, naturally, the size and severity of such flaws will be randomly distributed throughout the material. So the standard approach is to perform tests such as three or four point bend tests, then plot the results on a histogram, Fig **2**. With any luck it is then possible to fit a well known statistical distribution curve to this histogram which enables the designer to assess a mean strength and measure of scatter for his sample. Which curve applies, depends on the material and test conditions but where the strength is very much a function of the worst flaw present, then the Weibull distribution has been found most useful. This distribution curve, often called the weakest link model, was developed by Weibull in the 'thirties and applied successfully to problems such as the strength of steel used in Bofors guns.

The advantage of this approach is that it enables the designer to determine a survival probability for a given stress level. The gradient of the line is a measure of the scatter in strength of samples of the material. The gradient is called the Weibull Modulus, and materials with a high 'M' value have low scatter. Typically a good engineering ceramic would have M=10. Using such data the designer can now choose the safe survival probability for his component. For instance, by referring to a Weibull plot for the material the designer can read off the operating stress which will give him the 99.99% chance of survival.

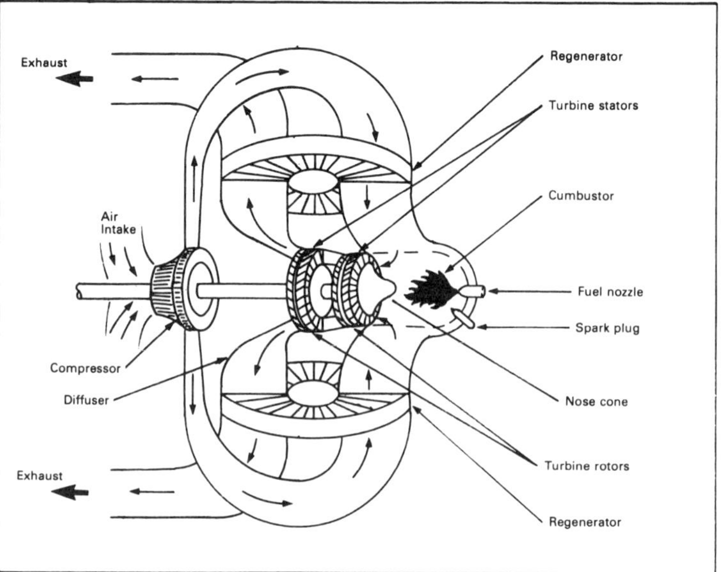

4 Schematic of Ford's ceramic turbine engine design

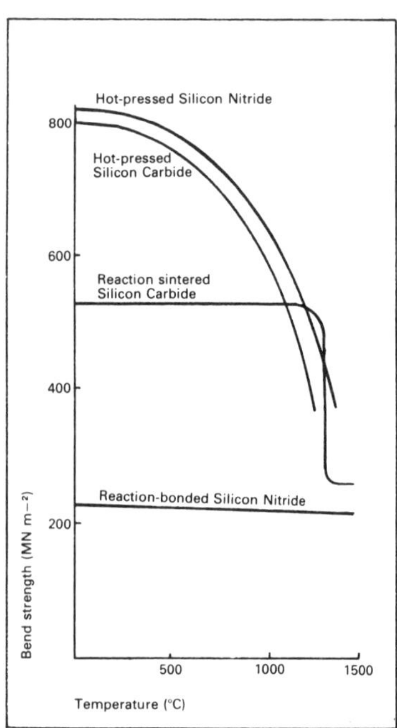

3 Strength versus temperature performance of ceramic materials fabricated with different processes

The choice of material

The last twenty years have seen a steady development of engineering ceramics. This has come about partly through aerospace requirements and partly for nuclear reactor components. Some of the most successful materials have been based on non-oxide ceramics such as borides, carbides, nitrides and silicides and it is from this range that two contenders were chosen for the ARPA programme. They are silicon nitride and silicon carbide. Now non-oxide ceramics such as these are usually fabricated in a very different way from conventional ceramic components. For instance a firing process such as that used for porcelain would be of little use for silicon carbide. Porcelain forms a glassy phase during firing which binds the other particles together on cooling. No such phase exists in silicon carbide so the grains of carbide can only be sintered together at high temperature under pressure where diffusion helps to bond them together. This is then called hot pressed silicon carbide and since most of the porosity disappears during hot pressing it is often called fully dense. The same can be done for silicon nitride.

There is, however, an alternative fabrication route for each which relies on a chemical reaction to produce the ceramic. In the case of silicon carbide a mixture of graphite and silicon carbide is compacted into a mould which has the geometry of the final component. The compaction is such that there is some residual porosity. The whole die is now immersed in molten silicon under vacuum which infiltrates the pores and reacts with the graphite to form silicon carbide. The final component contains residual silicon. Silicon nitride is formed in a slightly different way by heating compacted silicon powder in a nitrogen atmosphere until the reaction forming the nitride is complete. These products are then known as reaction sintered or reaction bonded ceramics. Produced in this way the ceramic has different properties from its hot pressed counterpart. Figure **3** shows how the strength of these materials behaves as the operating temperature is increased. Although the hot pressed forms are initially stronger, they lose strength rapidly as the temperature increases whereas the lower strength reaction sintered ceramics keep steady strength values to quite high temperatures. This gives the designer scope to employ hot pressed ceramics in relatively cool highly stressed parts and reaction sintered/bonded ceramics where high temperature performance is important.

A new concept in engines

Deciding on a criterion and a selection of possible materials is only part of the battle. From past experience designers saw that success was not to be had by simply replacing the hot parts of an existing engine with an exact ceramic copy. In fact Ford took a new look at the turbine engine and decided it had to be redesigned to suit these new materials. The result was an engine which employs ceramics in virtually all its hot stressed parts. Being a turbine for ground use it had to be regenerative, that is to say the exhaust gases must pass through a heat exchanger to heat up the incoming gases. Even the heat exchangers are made from ceramic material. Figure **4** shows a schematic diagram of the engine.

The first stage rotor, stator vanes and shroud, second stage rotor, stator vanes and shroud, turbine inlet nose cone, flame tubes as well as the regenerators or heat exchangers, are all made from ceramics.

To date silicon nitride has shown more promise than silicon carbide so this is where research effort is being channelled. Because silicon nitride exhibits this

Ceramics in turbine design

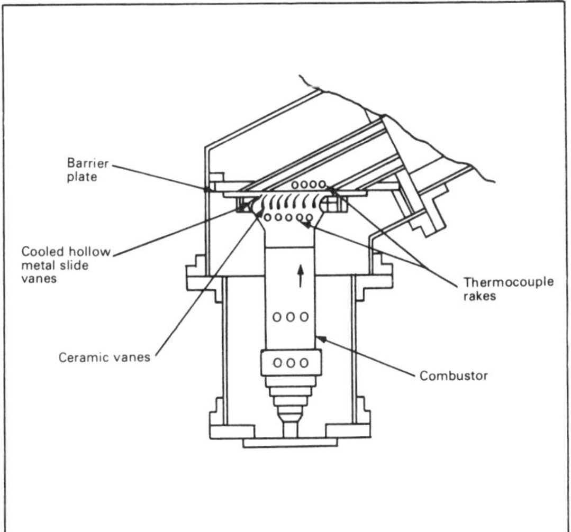

5 Static test rig for 1370°C test of ceramic vanes

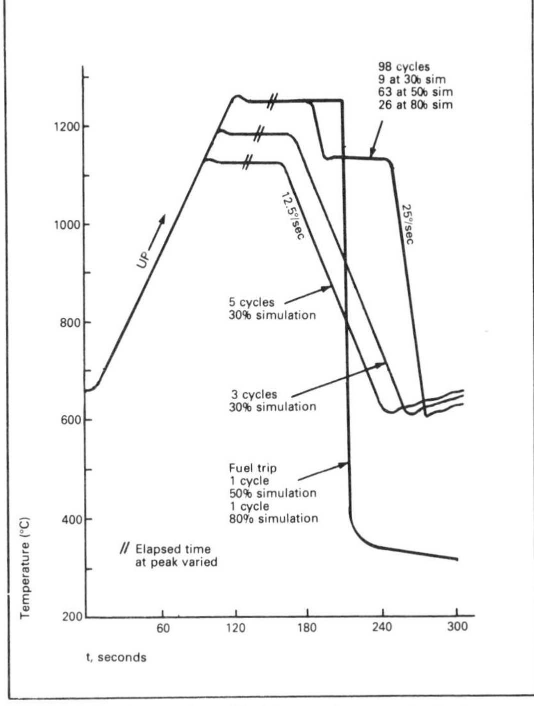

6 Typical cycles and cyclic history for ceramics tests

twofold set of properties, depending on manufacturing route, the designers have employed this to the best advantage in the case of the rotor. Now the rotor is more highly stressed in the disc near the shaft than at the blade tip; however, the blade operates at a much higher temperature than the disc. This led designers to the 'duo-density' approach in which the blades are manufactured from reaction-bonded silicon nitride, then fused on to a disc of the higher strength hot-pressed silicon nitride. The blades are integral with the blade ring, being moulded as one piece, so stress concentrations are reduced to a minimum. Most of the other parts are made from reaction-bonded silicon nitride. The whole engine is small for its output and so it has good potential as a vehicular power unit.

Assessing the engine's success

Perhaps the only way to be convinced of an engine's success is to see a fleet of vehicles leave the factory each day powered by that unit – especially if few are returned within the first six months of service. Such a commitment of either time or money was out of the question so it was necessary to assess the performance at each stage of the engine's development. The ultimate goal which the Ford designers aimed at was to demonstrate that the engine could run for 200 hours. Very few of the components did in fact give trouble over this operating period but the rotor, the most highly stressed part, did.

Rather than test the rotor in the engine and cause disastrous damage to the other parts it was decided to test rotors in a spin pit. This test involves spinning the rotor either to maximum operating revolutions and awaiting its destruction or increasing the speed until it disintegrates. Either test gives the designer useful data for redesign. A progression from this test was the hot spin pit where the rotor now spins at temperature. By carrying out such tests on a large number of rotors the results can be treated statistically by fitting a Weibull distribution to them, and so a survival probability can be determined for either a particular design or a certain batch of material. The results of these tests showed a particular sensitivity of survival probability depending on the conditions of fabrication. The rotor design employed now is quite satisfactory so discrepancies arise only from manufacture.

If the engine manufacturers have to test a sample of rotors every time a new batch of material is used, the engine no longer becomes a viable proposition so designers at Ford are now trying to correlate the results from simple bend tests made on samples of the ceramic with spin tests made on the same. If a reliable correlation is found, a simple, cheap quality control can be employed.

The large generating turbine

The story of the large engine is somewhat different. To begin with Westinghouse did not intend to design a revolutionary turbine engine entirely from ceramics. Their first aim was merely to replace the cooled stator vanes in an existing successful engine with a solid ceramic blade. Whilst this may not sound as adventurous as the Ford programme, it has unique and severe problems associated with it. The stator vane in the first stage of the engine is the component which suffers the brunt of the hot gas flow. Westinghouse were trying to design a stator which would operate at 1370°C for about 10 000 hours reliably. Added to this it was expected to withstand simulation tests such as shutdown when a fuel tripout occurs and a cold air blast hits the glowing vane thermally shocking it. So the conditions for the stator vane are very exacting. If this phase of the research proved successful the next step was development of a ceramic rotor.

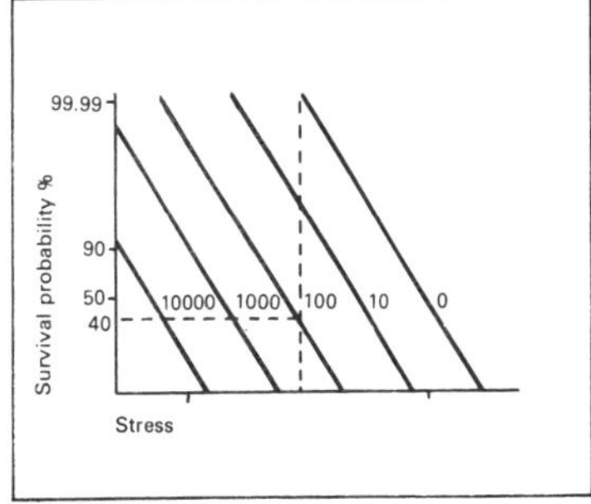

7 Strength probability time plot

Motivation again comes from combustion efficiency. Considering two existing turbine systems operating at 1150°C, an open cycle system operates at about 28% efficiency. A better system from the efficiency standpoint is the combined cycle system where the exhaust gases are subsequently used to generate steam for a steam turbine and thus boost efficiency to 45%.

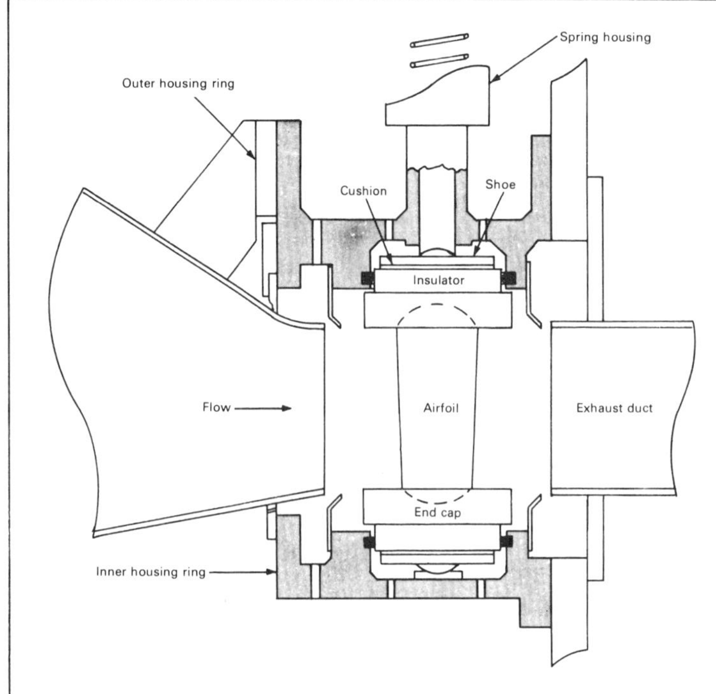

8 Ceramic three-piece stator vane design. A solid ceramic aerofoil is 'loosely' seated in female end caps allowing free movement under thermal and mechanical forces. Each end cap is supported by a single ceramic insulator, a cushion to distribute mechanical load and a metal shoe. The inner shoe is supported on two metal pivots, for three-point location stability, whilst the outer shoe is supported by a single spring-loaded metal pivot

These efficiencies can be increased to 33% for the simple cycle and 49% for the combined cycle if the inlet temperature can be boosted to 1370°C. Not only is efficiency boosted but also the power output. With an inlet temperature of 1150°C in the present engines the stator requires about 200°C of aircooling to drop the vane surface down to 900°C, which is the limit of operating temperature for super alloys over a 10 000 hour period – and this uses up engine power.

The cause of failure of super alloy vanes is usually corrosion and erosion. Ceramic vanes offer excellent corrosion/erosion resistance and would also allow low grade fuel to be burnt, such as gasified coal which would rapidly attack metal vanes. For a world fraught with power shortage and in no position to simply replace conventional power generation overnight by another system, these savings are very attractive.

The test programme

Unlike the Ford programme, Westinghouse did not intend to run the vane in an actual engine early in its development. Instead they chose simulation tests to be carried out in a hot gas test rig. Figure **5** shows the arrangement of blades in the rig and for most purposes this is a perfectly acceptable way of assessing the vane's capabilities. They chose to examine both hot pressed silicon carbide and nitride and also blades machined from billets of these materials.

A careful scrutiny of the operating conditions of generating turbines in 'peaking' situations enabled them to reproduce a good simulation of actual conditions which can be seen in Fig **6**. The severe fuel tripout condition is accompanied by several less severe cycles representing the ramp up of the engine during peak hours and its corresponding off peak shut down. These cycles are conditions of thermal shock, and superimposed on these transient stresses is a steady gas bending pressure which may operate for 10 000 hours. Long term operation may not appear to be a problem in a material which is extremely corrosion and erosion resistant but in fact the material does age in another and perhaps more catastrophic way.

Ceramics are prone to a phenomenon known as static fatigue or delayed fracture. This is a fatigue process associated with a stationary load rather than a cyclic one. Inside the material very small sub-critical cracks are growing steadily with time under load until they become critical. This means that the component can fail at a stress well below its fracture value if it is loaded long enough. Exactly what causes the cracks to grow is still a matter of much discussion but for certain materials it is generally thought that water vapour diffusing to the crack tip has an accelerating effect on the growth.

In recent years materials scientists have developed a statistical treatment of this problem and to a first approximation it is possible to incorporate this time dependence in the Weibull analysis. Results are obtained either from delayed fracture tests, where the specimen is loaded and then timed to fracture, or variable strain rate tests, in which a series of specimens is loaded at four or five different strain rates. These results can be plotted on a Weibull chart as shown in Fig **7** which is called a strength probability Time plot. Suppose the survival probability of the vane at start up under a given stress X is 99.99%. After 100 hours that vane will have a survival probability of 40% and if it has not already failed should be taken out of service instantly. This is a most useful approach but it only gives a first approximation, and so is treated with caution by the designer.

The vane geometry

As with the Ford engine a new design for the ceramic component was sought. One of the major problems facing the designers was the method of holding the vane. Most conventional holding methods set up contact stresses which were often sufficient to cause early failure.

The successful arrangement shown in Fig **8** employs a spring loaded pair of end caps specially profiled to accept the separate vane. Spring loading has the advantage of accommodating thermal expansion. The design work on this vane was aided greatly by finite element analysis of blade stresses and computer modelling of temperature profiles which, coupled with rig test feedback, has helped achieve this optimum vane arrangement.

The overall success

It is early yet to state firmly that the programme will be a total success but results from rig and engine tests are more than promising. Silicon nitride seems to be showing itself as the better material in most instances, and this was rather dramatically displayed during some recent rig testing of the large stator vane. Control of the flame tube went wrong and the inlet temperature quickly rose to about 1700°C. The engine was shut down immediately, thus causing tremendous thermal shock. Before testing a bank of four silicon carbide vanes were placed alongside four silicon nitride vanes.

On opening up the test chamber all that remained amidst the splatter of vaporized Hastalloy metal were four silicon nitride vanes and no trace of the others. The nitride vanes were still very much intact and in fact went back into test!

This design story has been made the theme of a case study in the Open University 'Materials under Stress' course. Calculations based on simple Weibull analysis show the student that the survival probability of the stator vane is gloomily low, and the question is posed "why, then, do the designers continue with this work?" In a television broadcast made for this course, some of the designers are interviewed so as to answer this question. After watching this programme, hopes that may have drained away from anyone interested in this project ought to be restored ●

Figs 13456 and 7 are reproduced from the Proceedings of the United States Second Army Materials Technology Conference.

Plastics as engineering materials Nylon 6

F J Parker*

Of all the engineering thermoplastics, nylon is the most widely used. Its combination of the following properties accounts for this:

- high degree of toughness
- wear resistance
- low friction
- hardness
- good chemical resistance
- low permeability to gases
- low volumetric cost.

These features make the material attractive to the design engineer for replacement of metal parts where it has benefits of cost, low weight, freedom from corrosion, the ability to consolidate parts and be easily moulded into strong intricate shapes. Also, having good electrical properties, it is widely used for electrical components where toughness and chemical resistance are additional requirements.

Modifications, such as glass-fibre reinforcement for higher stiffness and higher useful operating temperatures, acrylic modification for very high impact strength, and many other modifications to the basic polymer, make nylon a material with a wide variety of applications.

Nylon 6 is made by hydrolytic polymerisation of caprolactam which produces a crystalline polymer having a sharp melting point at about 215°C and a low viscosity.

Many of the properties of nylon 6, particularly its toughness, depend on the degree of crystallinity achieved in the parts produced and this in turn is affected by moulding conditions, post-moulding treatment and the crystallinity of nucleating agents in the polymer. An important additional advantage of nucleated materials is the ability to mould more rapidly. The polymer is hygroscopic and it is important to appreciate that absorption of water changes the physical properties (see Fig. **1**), indeed the maximum toughness of the material is not fully developed until mouldings have been conditioned by allowing absorption of water from the air or, more rapidly, in water (see below).

Properties

Of the mechanical properties of nylon 6 (Table 2), the impact strength is the most noteworthy. This property is enhanced by glass-fibre reinforcement when the property is measured dry but it should be noted that the normal conditioned state shows only small differences in impact strength between the reinforced and non-reinforced types. Glass-fibre reinforcement much improves the tensile and flexural strength of the material but this makes it less elastic as can be seen from the higher modulus values. An acrylic-modified grade is outstandingly different from the other grades having a greatly enhanced impact strength, although this is to some extent at the expense of flexural and tensile strength.

The change in properties from the dry to the conditioned state should be carefully noted and the conditioned values should be regarded as the normal ones (see section below on conditioning).

The electrical properties of nylon 6 (Table 3) are characterised by good insulation resistance and electric strength, although this is somewhat reduced by conditioning; the track resistance is also high. These properties make the material generally suitable for domestic electrical accessories (e.g. plugs, switches, etc.) at normal household voltages.

Nylon 6 mouldings may be used at temperatures up to 80°C, although as the temperature increases the modulus and strength decrease (see Figs. **2** and **3**). Glass-filled grades show a much smaller dependence on temperature as reflected in the values for the temperature of deflection under load at 1.8 MPa stress (see Table 4).

Above 80°C, standard grades are slowly embrittled by oxidation causing a decline in the mechanical properties. The same effect occurs on exposure to high levels of ultra-violet light. The rate at which this happens depends on the moulding thickness and environmental conditions but this effect can be substantially avoided with stabilised grades. Table 5 and Fig. **4** illustrate this with tests on 3 mm mouldings where the time to embrittlement is determined by observing when failure on a tensile strength test is breaking rather than yielding as is normally the case.

The chemical composition of nylon 6 itself confers a measure of flame retardancy (oxygen index of 25%), up-grading with flame-retardant additives can boost this performance.

Nylons are outstanding amongst thermoplastics for their resistance to many organic solvents, including petrol, oils and greases. Although some other organic solvents may swell nylon, especially at elevated temperatures, there are few reagents which will dissolve the polymer (e.g. concentrated sulphuric and formic acids and metacresol). Resistance to alkalis is good but acids will attack nylon, particularly if they are concentrated and at elevated temperatures.

Permeability to organic vapours and permanent gases, including oxygen and carbon dioxide is very low and nylon 6 is valuable where such barrier properties are necessary (Fig. **8**).

Processing

Nylon 6 can be both injection and extrusion moulded; blow moulding is also possible although the very low viscosity of the material makes this difficult for all but very small articles. The very much higher vis-

*Group Leader of Nylon & Polyester Moulding Materials Section, British Industrial Plastics Ltd

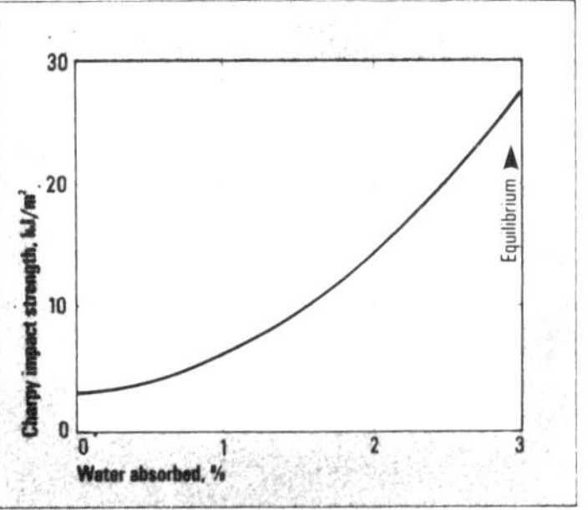

1 Effect of conditioning (absorption of water) on impact strength

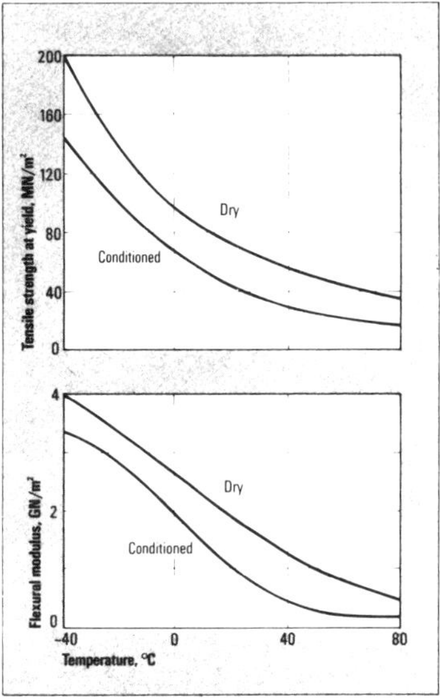

2 Effect of temperature on tensile strength
3 Effect of temperature on flexural modulus

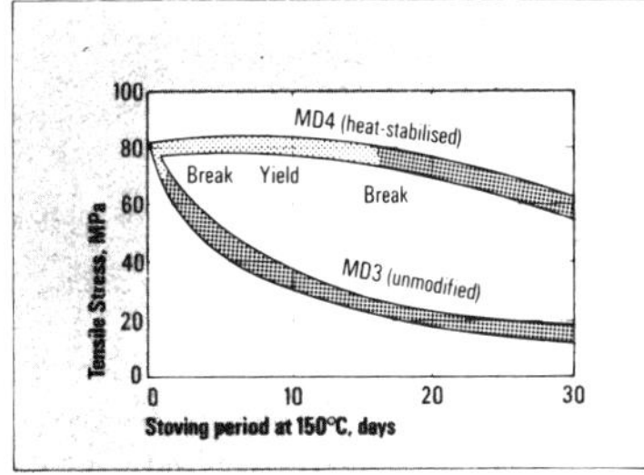

4 Thermal stability of unmodified and heat-stablised nylon 6
Unmodified nylon 6 (MD3) breaks without yielding after one day at 150°C, whereas heat-stabilised nylon 6 (MD4) shows a yield up to 16 days at 150°C

Plastics as engineering materials/Nylon 6

cosity of the acrylic-modified nylon makes this particularly suited to blow moulding.

The bulk of nylon 6 is injection moulded using standard thermoplastic injection-moulding machines operating with cylinder temperatures of 230 to 300°C. A nozzle with a shut-off valve must be used on the cylinder to prevent drooling which will otherwise happen with the very low viscosity of the material.

Non-nucleated grades when moulded at low mould temperatures (20 to 40°C) produce amorphous and translucent mouldings due to the low crystallinity achieved. Higher mould temperatures (70 to 100°C) allow the material to crystallise and this gives more opaque and stronger mouldings. Nucleated grades are less dependent on mould temperatures to achieve the desired crystalline condition and mould temperatures of 25 to 95°C can be used with these materials.

Glass-filled grades of nylon 6 need special consideration as it is important to fill the mould rapidly in order to obtain the best surface finish and gloss. This is achieved by using fairly high mould temperatures, 60 to 100°C, and using high injection pressures to reduce the filling time to the minimum.

The granular material supplied is hygroscopic and, if left exposed for long periods of time, will absorb sufficient moisture to cause defects on mouldings. It is important that partly used bags should be thoroughly resealed and that a lid should be used on the machine hopper. It is also important to realise that bags of nylon brought in from a cold-storage area should be allowed to warm up to the temperature of the moulding shop before opening, thereby preventing condensation of moisture on to the granules.

If a quantity of material is allowed to absorb moisture it should be redried before use. Ideally this should be done under vacuum at 80 to 100°C. An air oven may be used but this is slower and there is a danger of degradation and discolouration at temperatures over 80°C. Dry reground nylon can be incorporated at levels up to 20%.

An important benefit of using nucleated materials is the faster rate of production achieved owing to the very rapid set up of the polymer in the mould. This becomes increasingly important with thicker section mouldings where the cycle times may be greatly extended. For example, a heavy-section 10 cm diameter castor wheel required a minimum cycle time of 68 seconds using a standard non-nucleated grade of nylon 6, whereas a cycle time of about 52 seconds was sufficient for a standard nucleated grade and only 37 seconds for a very fast cycling nucleated grade. These differences can be more precisely and clearly illustrated by differential scanning calorimetry tests (see Fig. **5**) which clearly shows the more rapid on-set of crystallisation for

6

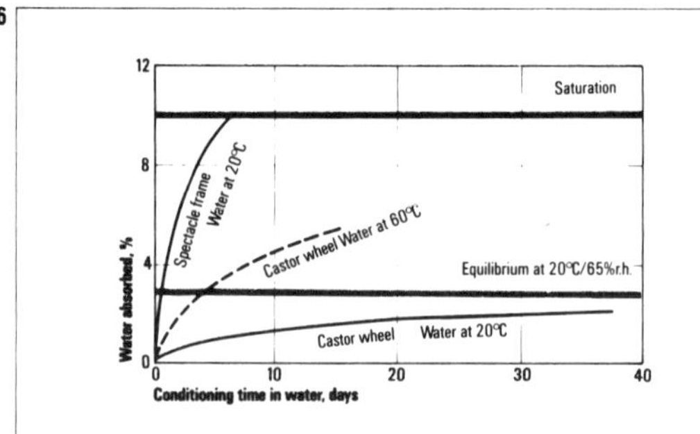

5 Comparison of crystallisation rates by differential scanning calorimetry

6 Water-absorption curves illustrating conditioning times for spectacle frame and a heavy-section castor wheel

7 Comparative volumetric costs

8 Comparison of nylon 6 with other thermoplastics

8

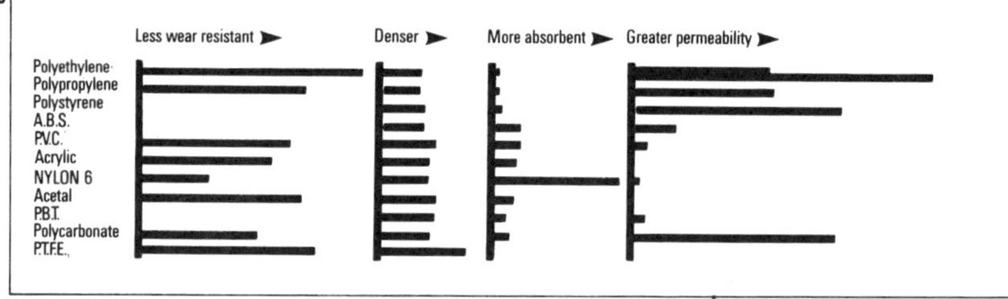

Gear Wheel: Nylon 6 chosen for self-lubricating properties, low friction and toughness
Car fan: Nylon 6 chosen for strength and resistance to petrol and oils
Fluorescent-light-bulb end cap: Nylon 6 chosen for insulating properties, toughness, ability to snap-fit and fast production rate

Industrial plug and socket: Nylon 6 chosen for insulating properties, toughness and resistance to harsh environments

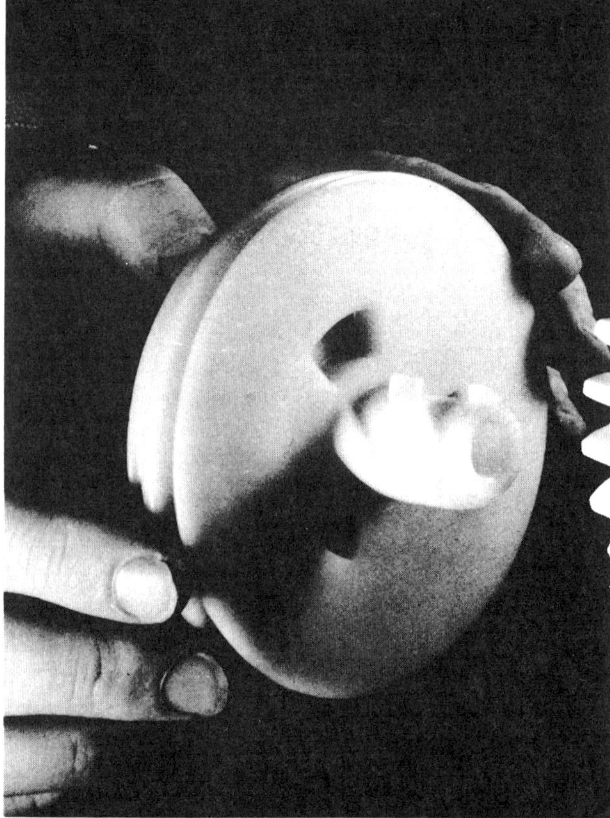

5
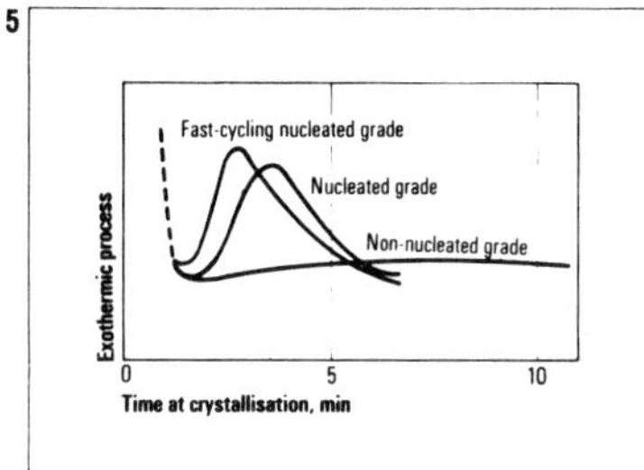

7
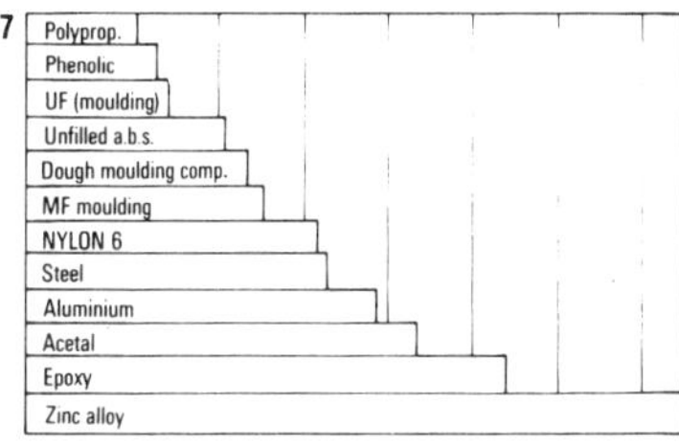

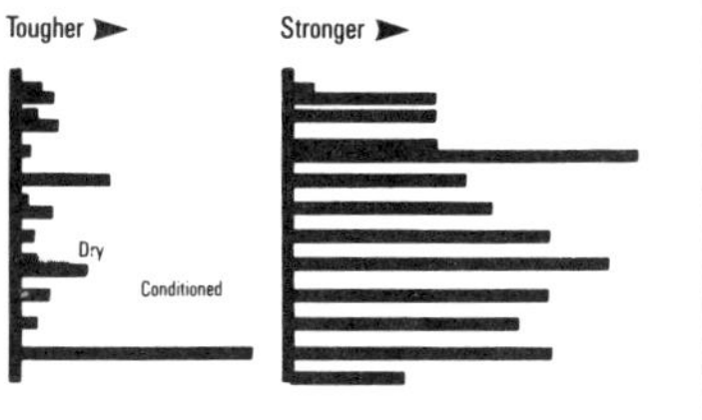

the nucleated grades. A further important benefit of such materials is their greater tolerance of widely varying moulding conditions and it is this flexibility in use which makes them most widely used for injection moulding.

Conditioning

As explained earlier, the process of conditioning, that is the absorption of moisture into the mouldings, is an essential part of the production of a satisfactory nylon 6 moulding. A moulding at equilibrium with normal atmospheres will have absorbed between 2½ and 3% of water. While this can be allowed to take place under normal conditions for very thin mouldings, the average part, and certainly much thicker ones, will condition too slowly in this way and it is, therefore, common practice to accelerate conditioning by immersion in water. The rate at which conditioning takes place will, of course, depend on the thickness of the moulding, its shape and the temperature of the water. The time required at room temperature is normally rather long and it is usual, therefore, to soak in water at 60 to 80°C until the necessary uptake of water has been achieved, Table 6 provides a guide to this process.

Mouldings are often conditioned for much shorter periods of time than this to achieve just a minimum degree of toughening, the remaining conditioning being left to absorption of moisture from the air.

Fig. **6** shows the rate of pick up of moisture in water at 20 and 60°C for two mouldings of very different thickness and this clearly illustrates the necessity of accelerated conditioning procedures for thick sections.

Finishing

It is normal for parts to be produced in one moulding operation, but mouldings can be machined when necessary. The relatively low modulus and thermal conductivity makes it necessary to pay particular attention to using sharp cutting tools, high cutting speeds, low feed rates and liberal amounts of cutting fluid to cool the part.

Nylon 6 can be welded by all the usual processes employed for thermoplastic materials and commercially obtainable adhesives can be used for adhesion to other materials.

Colouring by the introduction of pigments, either by the manufacturer or the processor, is the common form of colouring but the material may also be dyed using water-soluble dyes which are especially suited to nylon.

Application

Nylon 6 is widely used for many general applications in light engineering as an alternative to metals on account of its strength, toughness, resistance to wear, oil and solvents and heat as well as its freedom from corrosion. Gears and bearings take advantage of the low frictional and self-lubricating properties. The electrical industry uses nylon 6 for low-voltage accessories of many kinds, such as coil formers, plugs and sockets, switches, terminal blocks and power-tool housings. Automotive uses include handles, lock components, ball joints, bushes and bearings.

As the material can be steam sterilised, it is often used in the food industry for parts of machinery coming into contact with food and it is similarly used for many applications in the kitchen. Other home uses include parts for domestic applications such as vacuum cleaners and food mixers.

Marine applications of nylon 6 are widespread as it is ideal for boat fittings, previously made from metals, because of its resistance to salt-water corrosion ●

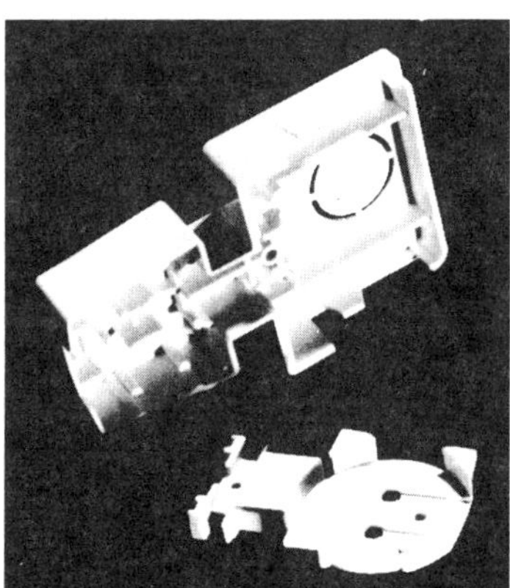

Type	Description	Use
MD1 HY1	Medium- and high-viscosity (no additives)	Extrusion moulding
LD2 MD2	Low- and medium-viscosity with lubricant for good mould release	Injection moulding of transparent parts
LD3 MD3	Low- and medium-viscosity with lubricant and nucleating agent	Injection moulding of crystalline parts. Fast cycling
MD3C	Medium-viscosity with lubricant and special nucleating agent	Very fast cycling. Fine crystalline structure
MD4	A heat- and light-stabilised form of MD3	Manufacture of parts exposed to high temperatures and ultra-violet light
MD6	MD3 with added moybdenum disulphide	Bearings and gears with very low friction
MD8	Medium-viscosity with flame-retardant additives	Parts exposed to fire risk
MDS1 MDS2	MD3 with 50% glass spheres MDS2 also contains heat and light stabilisers	Applications requiring improved rigidity and reduced shrinkage
MDF1 MDF2 MDF3	30% glass-fibre reinforced versions of MD3, MD4 and MD8	Applications requiring increased rigidity and strength, better mechanical performance at elevated temperatures. Reduced shrinkage
AC1	Acrylic-modified nylon 6 copolymer	Provides extreme impact strength and high viscosity. Suitable for blow moulding

Table 1 – Principal types of Beetle nylon 6 and their uses

		Non-nucleated LD2 MD2 HY2		Nucleated LD3 MD3 HY3	Nucleated MD4 MD6	Acrylic-modified AC1		Glass-sphere filled MDS1, 2		Glass-fibre filled MDF1, 2	
		A	B	A	B	A	B	A	B	A	B
Tensile strength	MPa	60	40	80	40	65	35	75	55	155	95
Elongation	%	5	40	5	30	5	100	4	6	6	7
Yield (Y) or break (B)		Y	Y	Y	Y	Y	Y	Y	Y	B	Y
Flexural strength	MPa	80	25	95	30	80	25	125	80	230	110
Flexural modulus	GPa	2.2	0.8	2.5	1.0	2.0	0.6	5.0	2.0	7.3	3.5
Charpy notched impact strength	kJ/m²	4	25	4	20 (HY3 30)	10	100	4	12	12	30

Table 2 – Mechanical properties: A – Dry specimens as moulded; B – Conditioned specimens to 2.5 to 3% water

	LD2, MD2, HY2,	LD3 MD3 HY3	Heat-stabilised MD4		Acrylic-modified AC1 MDS1, 2		Glass-sphere filled MDF1, 2		Glass-fibre filled	
	A	B	A	B	A	B	A	B	A	B
Electric strength at 23°C MV/m	14	10	14	6	12	7	10	10	11	9
Surface resistivity ($\log_{10}\Omega$)	15	12	15	11	15	13	15	9	12	12
Volume resistivity ($\log_{10}\Omega$cm)	15	13	15	11	15	13	15	11	15	12
Comparative tracking index	600	600	520	540	600	600	400	400	540	580
Loss tangent at 1 MHz	0.02	0.08	0.02	0.08	0.02	0.03	0.02	0.08	0.02	0.08
Relative permittivity at 1 MHz	3.5	4.0	3.5	4.0	3.5	4.0	4.0	4.0	4.0	4.0

Table 3 – Electrical properties: A – Dry specimens as moulded; B – Conditioned specimens to 2.5 to 3% water

	Acrylic modified	Non-nulcea-ated	Nucle-ated	Sphere-filled	Glass-filled
Dry	55	55	65	85	over 190
condi-tioned	55	55	60	65	180

Table 4 – Temperature of deflection under load at 1.8 MPa stress

Stoving temperature (°C)	Time to embrittlement (days)	
	MD3 (unstabilised)	MD4 (stabilised)
100	11	1500
110	5	770
130	2	380
150	1	25
170	Short	12

Table 5 – Stability to heat of standard and heat-stabilised grades

Thickness (mm)	Time to equilibrium in hours	
	20°C	60°C
1	8½	1½
2	35	6
3	80	13
5	220	35
10	820	150
15	1800	340

Table 6 – Conditioning time in water to equilibrium value at 20°C/65% relative humidity (i.e. 3% water)

LOOSE or TIGHT?

There can be very few people who have not suffered from things coming loose when they should have stayed tight. S H Grylls talks about some specific examples from his own experience and suggests improvements.

Every designer at some time or another has encountered nuts from 5BA to 3 in dia trying, not always successfully, to hold things together. This problem is not an unusual one and spills over into the domestic field where wheels fall off children's toys and off the family mower.

In a specific case of a child's toy the nut (a cap nut) bottomed and, on the mower, the nut met a shoulder. In both cases there was no way of holding the shaft with an adequate torque until the nut was tightened to a sensible figure. The modification of the mower was simply, **1**, to add two flats to each axle head to hold the shaft and applying the correct nut torque.

Generally, the use of set screws, bolts and nuts falls into two categories. In the first category the thread is used to stop things falling apart in such items as the blades of a pair of scissors or the wheels on the child's toy as mentioned above. In the second category, the threads are used to hold things tightly together, preventing any relative movement. Examples are the ring set screws holding the crown wheel to a vehicle differential or the retention of the steering lever to a stub axle. To a purist there is a third category which is difficult to describe in general terms but of which a particular example is the nut securing a splined coupling to a splined shaft.

A nut which prevents things falling apart must be locked or tightened against a shoulder so that any loosening torque in use is far less than the initial tightening torque. A nut which holds things tightly together needs no locking if it is man enough for its job and really prevents relative motion of the parts it holds together. A paradox is to be found in the usual insistence on locking big end bolts while never even considering doing so for bolts securing cylinder heads and sumps. In theory both of these are just as likely to come loose.

Common objects with a fascinating variation in design are secateurs, garden shears and scissors. The blades have to be held together with a small preload in the case of scissors, or minimal slack for secateurs; the preload must be easily adjustable; the assembly must not alter its adjustment in use, ie when the blades undergo relative motion. A headed pin, some adjustment washers and a split pin could fulfil the requirements.

Scissors

Most scissors are, however, held together as shown in **2**, the blade furthest from the bolt head being threaded and the end of the bolt riveted over. In many scissors, (due to poor craftsmanship the designer will say!) all three essential conditions are unfulfilled. The same design can be seen on flower scissors where the blades tighten in use. Very skilled adjustment and re-riveting is required to make them work. The screw-head, counter-sunk for appearance, acts as a cone clutch. As usual the frictional coefficient μ seems to be high where one hopes it to be low and vice-versa.

Some secateurs and long handled pruners are held together by a hexagon headed bolt screwed into the farther blade and locked by a nut, **8**. As this nut is tightened, the thread in the blade stretches and the adjustment of preload between the blades alters. Trial and error are required to set this adjustment but the assembly does stay tight if a spanner is used on the bolt head while tightening the nut to a sensible torque; **3**, shows a better arrangement. The inner nut is tightened to give the correct preload or position, the outer nut then being tightened while holding the inner nut. The part of the bolt passing through the two blades is no longer stretched during locking and the initial adjustment is not altered.

Another arrangement is to provide the preload by a Belville washer, a large nut being locked in any position by one blade of a multi-tabbed washer. Since the Belville washer is not flat, the blades can be forced apart when cutting and a material like string jams between them.

From examination in the garden shed, there appear to be eight different ways of holding two blades together. Scissors and long bladed shears depend for their pre-load on the curvature (in side view) of the blades. Secateurs depend on the axial loading of the retaining pin or bolt. The blades must not tilt on the pin and need a good 'wheelbase'.

There are of course two kinds of secateurs, the scissors type and the anvil type. Strangely enough, the latter, although not requiring such good control in the direction of scissor blade separation, have in every case greater rigidity than their rivals. This is because the U-section has a good wheelbase. It may not be convenient to use this construction for scissor secateurs because of the selective assembly or washering needed to position the blades. The design of **12** is worthy of consideration.

The many arrangements are easier to understand by reference to **4-12.**

Consider next the front hub of a car. The bearing arrangement varies between two ball bearings, a parallel roller bearing and a ball bearing and two taper roller bearings. The bearings can be slightly preloaded or fitted just slack. One certain thing is that the inner race of the outer bearing will creep and will exert a considerable torque on the securing nut either forwards or backwards. This basic set up is essentially as shown in **13,** bottom half, but a method is required

Continued on page 472

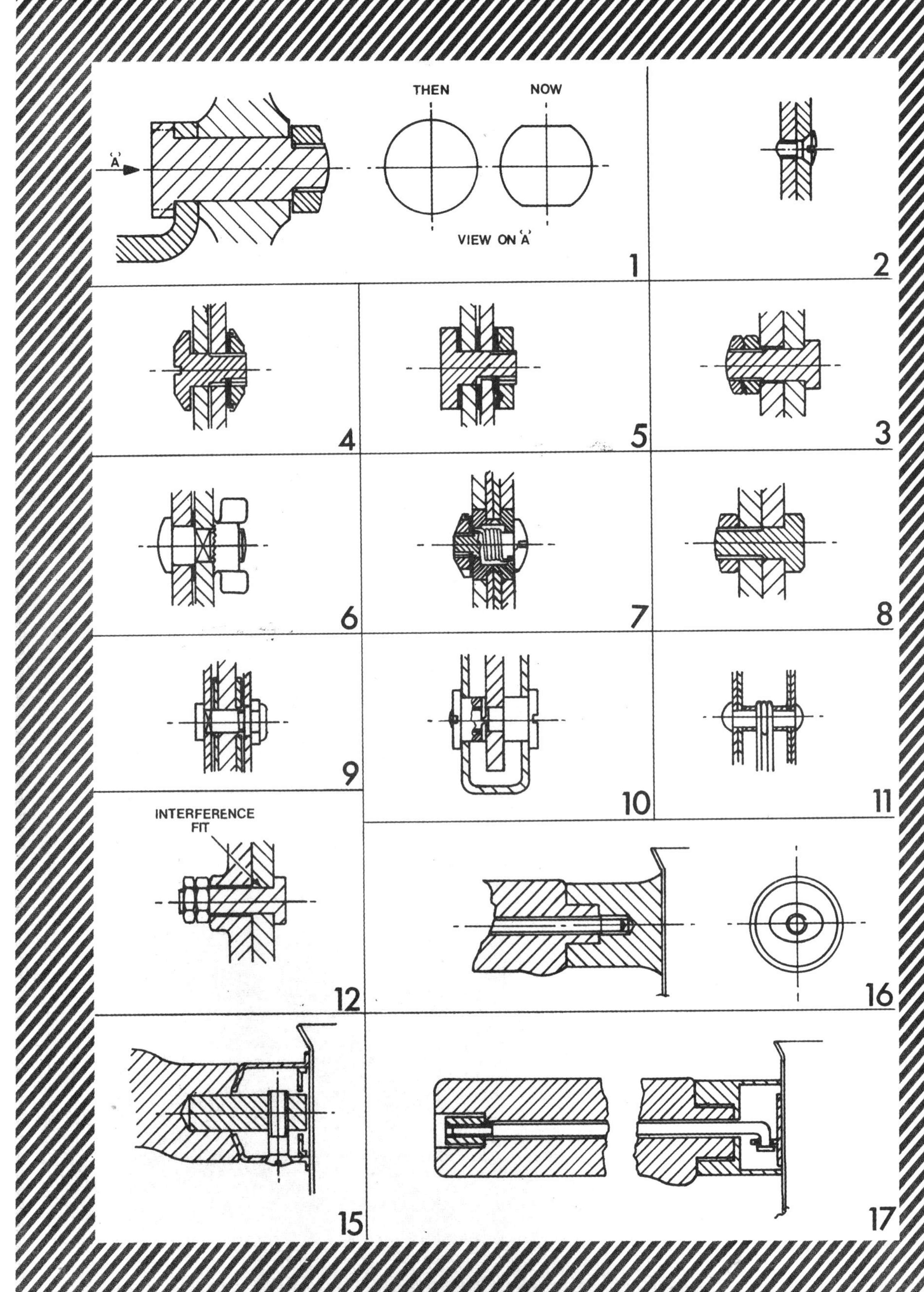

THEN
NOW
'A'
VIEW ON 'A'
1
2
4
5
3
6
7
8
9
10
11
INTERFERENCE FIT
12
16
15
17

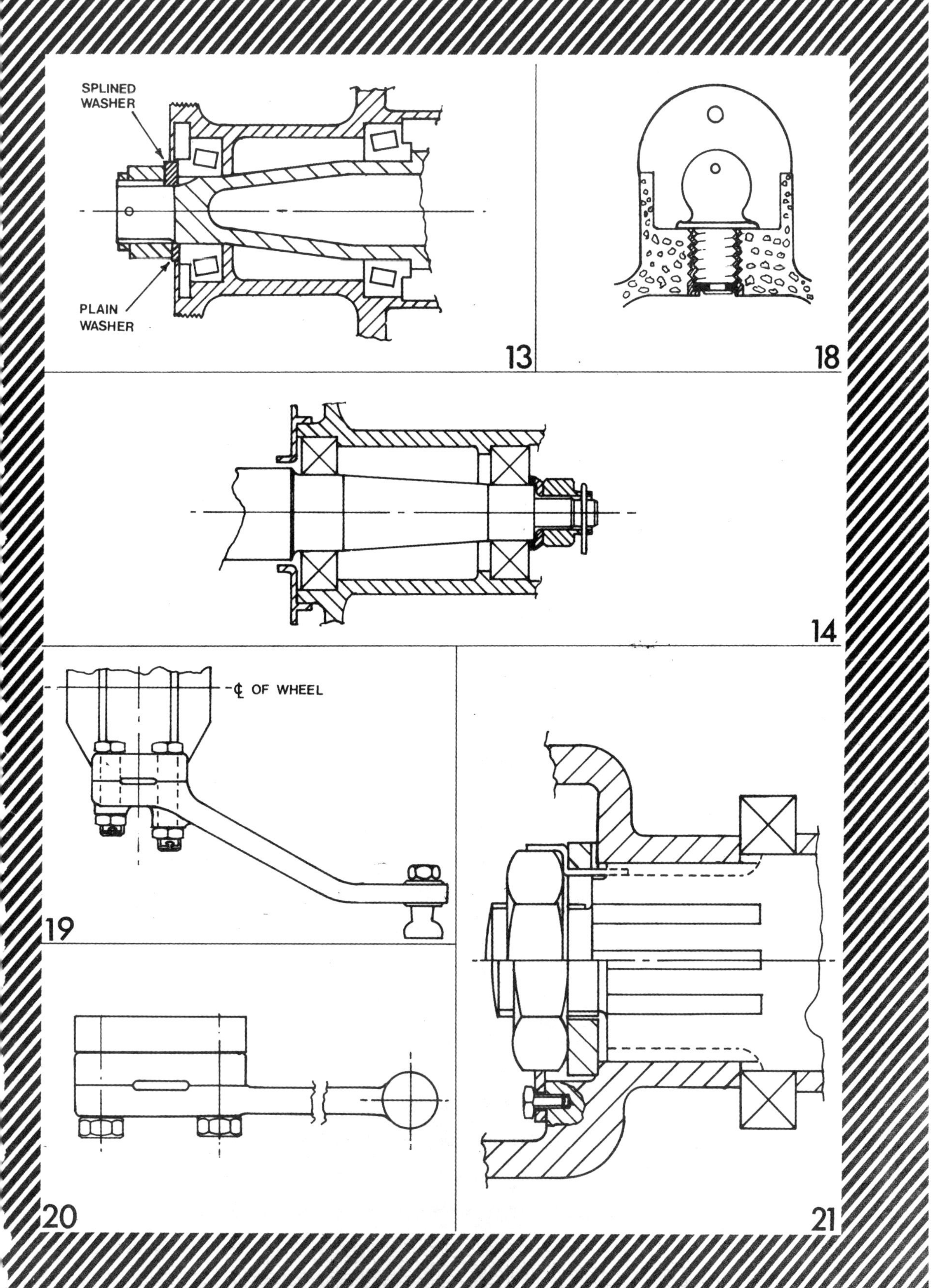
SPLINED
WASHER
PLAIN
WASHER
13
18
14
℄ OF WHEEL
19
20
21

Loose or tight

of relieving the nut of any torque from the outer bearing's inner race.

A common arrangement is to insert a thick splined washer between the race and the nut, **13,** upper. If the bearings are slack, so is the nut, a large split pin preventing it from falling off.

An improvement would be to tighten the nut against a shoulder; an adjustment washer can be varied in thickness to achieve the correct slack, **14.** Whether the nut needs to be split pinned is no longer a matter of engineering during ordinary motoring, but it would help to counter the effect of exceptional overloads when sliding bodily sideways into an obstacle.

Saucepan handle

Reverting to the similarity between domestic designs and those in much heavier engineering, in the kitchen there are three different ways of retaining saucepan handles. Two come loose and the third stays tight for the same reason as the finally successful industrial torque converter output shaft coupling, described later.

The worst saucepan handle, **15,** is loosely located in the pressed bracket riveted to the pan and is held by a short grub screw struggling against loads totally beyond its capabilities. Alone it tries to counter twisting and bending loads on an impossibly short base. All the important fits are slack. The second design, **17,** is an improvement in that a long screw, the whole length of the handle, contains considerable strain energy and hence can absorb some fretting wear or compression. If, as in the third saucepan, **16,** the handle had been an interference fit of reasonable length in the pan and of non-circular section, the long bolt is hardly required at all.

Not really connected with screw threads but exemplifying correct engineering are the stopper of a hot water bottle and the car doors of a 1953 Pontiac. Although the engineering of the Pontiac is no longer used, at least hot water bottles can carry on the tradition if correctly dimensioned.

The seal of the water bottle is at the bottom of the thread. It, like the rubbers on the Pontiac doors, is of low rating but cannot be overloaded because the much higher rated stop at the top of the stopper comes into play at the correct deformation of the seal proper, **18.** The Pontiac doors had very large flexible seals which still worked after 70 000 miles and whose compression was limited by two hard rubber stops on each door.

Items that must be held tightly together and must not fret and wear at the mating surfaces are in many ways easier to engineer. The attachment by a ring of set screws or studs of a crown wheel to its differential housing is one such assembly. The driving torque between the two parts should be taken entirely by friction since to rely on the shear strength of the bolts admits to accepting relative motion which can reverse in direction. The maximum torque on the crown wheel is engine torque × 1st gear ratio × axle ratio × the excess capacity of any clutch in the train. The latter can be 2 or 3 to 1. The mating surfaces of the crown wheel and differential are usually smooth but of course have no machining marks across the direction of the relative motion. It has not proved possible in my experience to rely on a μ in excess of 0·1 for an assembly tightened once and then hidden for all time. If retightened after some initial fretting μ appears to reach at least 0·25 and half as many set screws would suffice.

The force required to slide one object or another can be represented in a simple graphic form. Movement starts right away and for steel on steel there is a sharp kink in the curve at $\mu = 0{\cdot}06$*.

Steering attachment

The steering lever to stub axle attachment at the front of a car is another example of clamping which must not shift. Two arrangements are usual, in one, vertical set screws ensure that steering is by friction. In the other, horizontal set screws steer in tension. **19** and **20** show these two arrangements. In the first, the shear strength of the bolts would come into play during an exceptional shock overload. The size of screw should be chosen so that the maximum load not in collision requires μ to be no larger than 0·1. How to calculate this is not easy, the rigidity in bending of the parts being difficult to take into account.

The second case seems to function with smaller bolts, although practice does not bear this out. Assuming the mating surfaces are made with a cut-out so that no possibility of clamping convex faces arises, steering in one direction loads the nearer bolt and in the other direction, the farther bolt. The bolts can be, and sometimes are, of different sizes. The problem should be the same as connecting rod bolts which give no trouble if their initial tension exceeds the maximum applied load. Bending of the lever now puts an unequal load on the bolt head, a very important and rather incalculable effect. In a particular case all early experimental testing was successfully done with unlocked set screws but later stainless steel tab lock washers were added. Trouble soon arose due to the very small μ of the lock washers whose tabs gracefully moved out of the way. A choice then existed of explaining to everyone involved that the lock washers should either be thrown away or a lock devised which did no harm. The latter solution was adopted!

In retaining a splined or taper coupling on a shaft the coupling is there to transmit torque but may have to endure end loads which are sometimes truly axial but more often inclined particularly where a Hooke's joint is bolted to it.

The coupling is almost certainly retained by a large nut or a long set screw passing through it and the different modus operandi include:— Nut used to end load all the items on the shaft. Nut tightened against a shoulder allowing slack in the assembly. Nut locked to the shaft. Nut locked to the flange. Coupling a sliding fit. Coupling tight.

The two halves in **21** show all the possibilities. Torque reversal sufficient to cause oscillation of the coupling on its splines in (1) exerts a large torque on the nut and in (2) exerts a torque on the nut when the coupling is pulled rearwards.

If the nut is locked by a washer to the flange it might be concluded that both will rock a little together and not give trouble. Wear on the thread will however probably put a finite life on the assembly. If the coupling is a drive fit with sufficient interference on the splines neither it nor the nut will move. If the splines cannot be tight then an enormous nut fully tightened against an adequate shoulder is the best choice. Brute force will eventually win.

If the maximum load on any nut or set screw head were truly axial and never larger than that due to tightening, life would be easier but much less interesting than reality. By aiming for the simple condition with brains or brute force trouble can be avoided. A split pin is a good form of inspection; its fitting probably means that the nut has just been correctly tightened. A dab of paint has proved equally effective and very much less costly●

*This information was provided by F T Barwell in a paper entitled 'Surface contact in Theory and Practice' read to the Institution of Mechanical Engineers in 1961. It may explain, however, why crown wheels, etc, will fret if clamped so that the classically accepted figure of μ is relied upon.

Automobile engineering – a spectator's view

Materials problems

Rubber

By and large, rubber and elastomer elements have too short a service life. It would add such a minute fraction to the total product cost as to be not noticeable, yet it could substantially reduce operating costs and increase brand loyalty and regain some of the sales lost to overseas competition. There is no excuse for having these parts harden, age-crack, extrude or oil-swell after a couple of years of normal service, let alone fail by the rubber tearing or by breaking of the rubber-to-metal bond. There should be no cause for purchasers still to complain of badly sealing doors which whistle at speed, seals which come away from door or boot frames, water ingress past windscreen surrounds, a boot which fills with water, water hoses which burst or age-crack, constant velocity joint boots which fail to keep oil and grease in and dirt out, oil leaks past seals.

The every-day operating environment of heat, ozone and oil mist, road grit, slush and rainwater is not stipulated in many of the specifications against which the specialist firms supply. Each one of these factors tends to affect adversely not only the performance of these rubber elements but also much more costly units which depend on the proper performance of these rubber elements.

For instance, the type of rubber selected for exhaust system mounts is far from ideal for tensile loading. Failure of, for example, a simple rubber ring or a rubber and fabric strap is quite common. It may cost relatively little to replace it, but, by the time its rupture or disappearance is noticed, the effect on the life of a long run of exhaust piping may be considerable – and that is a much more costly affair. It may also prejudice a purchase of another car from the same manufacturer.

Rubber gaiters on rack and pinion steering and front wheel drive universal joints all too often suffer from three defects:

1. The metal clips securing them corrode away too rapidly – the factory-provided anti-rust protection is too short lived and there are no easy methods of improving this *in situ*. So they should be replaced periodically as a matter of preventive maintenance – but rarely are they renewed in time to save a major bill.
2. The rubber compound is not resistant enough to the real life conditions it has to operate in; salt, mud, slush, oil and solvents, ozone and even the occasional spill of anti-freeze solution.
3. The sharp contoured rubber convolutions tend to fail in fatigue, thereby allowing the lubricant to escape and dirt and grit to enter – net result – a costly repair.

The poor performance of one of our small sports car's rubber-bonded engine mounts appears to be directly related to oil-fouling deterioration, resulting in softening and eventual collapse. Undoubtedly the worst to suffer from this are the Triumph 13/60 and its Vitesse derivatives, where a failure to give satisfactory service reached nearly 1 in 4 on six to eight year-old cars. The Triumph Spitfire fared little better, yet the cure, more oil-tight engines and dipping the finished components in an oil-resistant rubber paint, is cheap enough.

Anti-roll bar bushes which perish, particularly on MacPherson strut front suspension systems tend to result in needless tyre wear.

A modern car relies heavily on compliance or controlled flexibility and built-in vibration and sound insulation, provided by mounts and bushes using rubber in compression and/or shear. A change of these properties, due to whatever cause, is bound to affect other units and ride comfort adversely. It is a relatively simple and cheap alteration to the material specification of rubber mounts, such as dipping them in neoprene compound which makes them oil resistant.

It seems strange therefore that by-and-large UK motor manufacturers do not demand of their suppliers that the anti-vibration mounts and similar rubber elements are made in material which can cope with *all* the environmental conditions, among them oil fouling and heat, which are, after all, to be found close to many power units, gearboxes and final drives.

Oil tightness of engines and transmissions of our European, Japanese and North American competitors has been achieved to a much higher standard than in many UK counterparts. Even so it is still virtually impossible to ensure, over a long period of service, total absence of oil mist in the engine compartment or some seepage of automatic transmission fluid.

The best of our UK specialist sealing strip and oil seal manufacturers are amongst the most competent and competitive in the world and regularly supply Continental vehicle manufacturers. But there are others who are unable or unwilling to do better than supply 'best black rubber' – for that may still be all that the detail drawing calls for. This is particularly so when it comes to some cut-price and/or some spurious parts suppliers. In the interest of the survival of the UK motor industry this practice, which is false economy, should cease.

'. . . making hoses fulfill a role they should not be called upon to perform, i.e. help limit the engine rock at idling or due to a sudden change in engine torque'.

The effect of water, slush and rock salt solutions on suspension and subframe rubber-metal bushes may be worth another look. Similarly telescopic damper rubber-bushes suffer premature age degradation. Cannot these units be made 'fitted for life' components? Our investigations on thousands of cars over two years old shows that age degradation can be so severe in some models as to cause rubber-to-metal bond failure after about two to three years, or severe swelling and substantial change in dynamic modulus, even rupture or the rubber itself, in others. Improved elastomer sealing strips should eliminate such irritating features as loosely dangling door seals or corroded-away boot seal retaining clips.

The latest type of door and window surrounds available are of elastomer materials (e.g. e.p.d.m.) which, though not cheap, can and do resist the effects of ozone, oxidation, and ultraviolet degradation ageing, and fluctuations of extremes of temperature and still retain their built-in resilience for many years.

The problem of bonding e.p.d.m. to other plastics and foam rubber has fairly recently been resolved and this makes it possible to overcome those persistent complaints of wind noise past door seals, body rattles and squeaks. In a recent sample of 1500 new cars (these were of the 10 leading UK models at the time only) some 30% were found on delivery to be in need of rectification for such faults. Door and lock re-alignment by the franchised dealer cured only a small fraction. About 350 were still noisy. Many of these would have been curable by having a more resilient type door surround sealing strip, which requires little effort to fit on the assembly line and, once pushed over the body flange, is difficult to loosen or remove.

It is easy to blame a car manufacturer's material specification engineer for being tardy in adopting an e.p.d.m. or comparable material – but these materials do not easily bond to each other or most other compound materials, whereas the s.b.r or modified s.b.r. and n.r. materials, which have been used in cars for decades with some success, do so quite readily.

Burst water hoses are still a major cause of breakdowns. Both factory-fitted and those cheap and nasty ones supplied by the corner garage tend to fail prematurely. In part this is due to a paring down of materials specification of hose and clip, and in part due to making hoses fulfill a role they should not be called upon to perform, i.e. help limit the engine rock at idling or due to a sudden change in engine torque.

Often the excuse given is that there is insufficient justification (based on warranty claims) for a raising of material standards. This misses the point that most failures occur after the warranty has expired, but well before competitive cars are immobilised by such a silly component failure. Under-bonnet temperatures are steadily rising. This surely calls for a change of rubber specification of all parts affected thereby: hoses, ignition leads, engine and transmission mounts.

'If only the water hoses and h.t. leads on some of your quality cars were as good as those of Mercedes Benz' is a cry from the heart of many a well wisher in Europe.

The majority of cars have reasonably long-life engines and gear boxes. Why then do so many suffer from 'entrail troubles' as a young friend of mine so graphically described it when the rubber boot lid seal of her car came away.

I do not recommend her drastic measure just to rip it off altogether. The choice of a rubber, which swells or stretches with age, cracks with long term exposure to air and/or ozone, the poor quality of fasteners from cheap staples to push-on clips with limited anti-rust protection are not in the best tradition of craftsmanship. Nor should they be fitted so haphazardly that the nearside door seal is an inch shorter than the opposite one when you remove them and lay them side by side.

Nylon

In the late 1950s several manufacturers discovered that the greasing of knuckle joints on steering and some suspension linkages could be totally eliminated – at a draughtman's stroke – by using nylon, preferably impregnated with additional solid lubricants. They were the first generation of fitted-for-life components – regrettably a very short one in far too many instances.

For, as in the case of early plastic macs, the material selected was badly chosen, thereby giving plastics a bad name. The nylon was hygroscopic and in our moist climate, this brought about significant changes of dimensions sufficient to cause tightening up on the metal ball, accelerating the rate of wear of the steel to the point of eventual danger of separation of the joint.

The deposition of rock salt on our roads for snow clearance also affected the nylon and the protective dust covers made of rubber. There is still room for improvement in specifying the correct sort of material to guarantee a service life of 10 to 12 years. It means that they must be able to resist *all* the environmental conditions from dry concrete dust to salt-rich slush.

Surprisingly we still find regular runs of 'thin case' trouble on several well known makes, particularly French. It is a case of insufficient control in material and heat treatment processes. Where it occured the hardened case was found to be much too thin to support the repetitive contact loads. After pitting fatigue the case was then worn away in a relatively short time. Replacement costs £100 to £180 to the owner – cost saving to the manufacturer a few pence only.

Specifications

The process of updating material specifications is a never ending one and has resulted in substantial changes of design criteria and test procedures. Amongst the large scale manufacturers VW lead the field. In the UK Ford followed and now demand functional performance testing of the finished product rather than simple chemical and mechanical laboratory tests.

Chrysler has been quick to change after field failure reports reached them. Over the past 12-18 months Vauxhall has been actively improving it's quality by detail design and material specification changes.

It will take a year or two before the whole of the Leyland Car organisation has caught up by harmonising its design and purchasing activities to bring material specification and testing methods up to the best that Europe and the rest of the Western World has to offer.

Acting on information on what went wrong in the field – often well after the car warranty expires – is most important. The average age of cars in daily circulation in the UK is 5½ years. Some models have been around with the same basic design and material specification shortcomings for 12 years or more. There still remains the problem of lack of oil tightness on engine, gearbox and final drive, or unsatisfactory drive shaft couplings due to oil fouling, water ingress past sliding windows, and leaks of road spray past convoluted gaiters on steering racks.

'In terms of reliability and performance there has been progress after all and there is no reason to suppose it will stop in the foreseeable future'.

There is little room for complacency. Our motor industry depends to such a large extent on merely assembling bought-in components. A fresh look at the environment in which such components and assemblies have to perform, and better co-operation between material engineers, purchasing departments, designers, production engineers and quality inspectors is needed. More realistic component testing and re-writing of specifications away from 'material' towards 'performance standards' in co-operation with the leading elastomer manufacturers and raw material suppliers are also overdue.

At one time some Japanese materials were poor, and East European ones even worse, but this no longer holds true for Japan. Also a number of Iron Curtain country manufacturers are not so many years behind in technology as to give the UK manufacturers a great deal of time to put their house in order before they too will compete on a world-wide basis.

Conclusion

We often hear derogatory comments about the quality of current design and manufacture coupled to a nostalgic hankering after vintage models of cars and aircraft. A visit to any Transport Museum or a careful study of designs of some 25 to 30 years ago will go a long way to dispel the myth of 'The Good Old Days'. By today's standards many of the exhibits are below par. In terms of reliability and performance there has been progress after all and there is no reason to suppose it will stop in the foreseeable future●

Bibliography

(1) There are many British Standards relevant to the matter of material selection. Many useful standards will be found listed in the Sectional Lists for Iron and Steel (SL24), Non-ferrous metals (SL19), and Leather, plastics, rubber (SL12). These lists are available free of charge from British Standards Institution Sales Dept, 101 Pentonville Rd, London.

(2) Engineering Design Guides, published for the Design Council, the British Standards Institution and the Council of Engineering Institutions by Oxford University Press. Of the titles so far published the following are of particular relevance to the matter of material selection:

No. 19 The selection and use of thermoplastics.
No. 26 Materials for low-temperature use.
No. 27 Introduction to materials science.
No. 28 Materials for high-temperature use.
No. 29 The selection of materials.

(3) The Open University, Technology: A third level course, Materials Processing (T352). Unit 10 Casting Processes. Unit 11 Polymer Processing. Unit 12 Working Solids. Unit 13 Joining and Shaping.

A checklist of books in the Butterworths Technician Series

MATHEMATICS FOR TECHNICIANS 1

FRANK TABBERER, Chichester College of Technology

This is an introduction to mathematics for the student technician, intended especially to cover mathematics at level one in TEC courses (core unit U75/005). The presentation will create an interest in the subject particularly for those students who have previously found maths a stumbling block. There are frequent examples and exercises, with a summary and revision exercise at the end of each chapter.

CONTENTS: Manipulating numbers. Calculations. Algebra. Graphs and mappings. Statistics. Geometry. Trigonometry.

192 pages May 1978 0 408 00326 X

MATHEMATICS FOR TECHNICIANS 2

FRANK TABBERER, Chichester College of Technology

This covers mathematics at level two in TEC courses (units U75/012 and either U75/038 or U75/039), for those who have completed (or gained exemption from) the work in *Mathematics for Technicians 1*. It includes the alternative schemes of work allowed in the second stage of level two. The clear presentation and systems of examples and exercises, similar to those in the first volume, will enable students to gain a real grasp of the subject.

CONTENTS: Trigonometry (1). Areas and volumes. Statistics (1). Graphs. Trigonometry (2). Equations and graphs. Mensuration. Statistics (2). Introduction to calculus.

156 pages September 1978 0 408 00371 5

PHYSICAL SCIENCE FOR TECHNICIANS 1

R. McMULLAN, Willesden College of Technology

This is intended for students studying the Physical Science level one unit of programmes leading to TEC certificates and diplomas. The text meets the requirements of the standard TEC syllabus for physical science (unit U75/004), a core unit of courses in building, civil engineering, electrical engineering and mechanical engineering. Attention has been paid to the visual presentation of the text, which is illustrated with diagrams and examples. Important concepts and formulae are clearly highlighted as an aid to learning and revision.

CONTENTS: Introduction. Fundamentals. Force and materials. Structure of matter. Work, energy, power. Heat. Waves. Electricity. Force and motion. Forces at rest. Pressure and fluids. Chemical reactions. Light.

96 pages May 1978 0 408 00332 4

ELECTRICAL DRAWING FOR TECHNICIANS 1

FRANK LINSLEY, Reading College of Technology

Written especially to cover the objectives of TEC unit U75/011, an essential unit for first-year electrical engineering, electronics and telecommunications students, this book explains the principles of general engineering drawing, electrical and electronics drawing, and product design. Self-assessment exercises (with model answers) are included to test the student's ability to meet the unit's specific objectives.

CONTENTS: The rules of drawing. Orthographic projection. Engineering symbols and abbreviations. Auxiliary views and surface development. Pictorial projection. Dimensioning. Electrical symbols and abbreviations. Electrical diagrams. Circuit diagrams from equipment. Product design. Product manufacture. The buyer's choice.

96 pages June 1979 0 408 00417 7

TELECOMMUNICATIONS FOR TECHNICIANS 1

G. L. DANIELSON and R. S. WALKER, South London College, Norwood

This book covers the Telecommunications Systems 1 TEC unit (U76/007) and is intended for first-year students who later on will specialise in either line transmission or radio systems or proceed to a general interest in electronics. It will also be useful to anyone looking for an introduction to the subject of telecommunications.

CONTENTS: Transmission of information. Modulation and demodulation. Radio waves and their uses. The cathode ray tube. Television. Radar. Use of radio for navigation. Telephony and telegraphy. Routing a signal. Transmission of speech by landline. Data communication.

112 pages May 1979 0 408 00352 9

ELECTRICAL AND ELECTRONIC PRINCIPLES 2

I. R. SINCLAIR, Braintree College of Further Education

The book covers the aims and objectives of TEC unit U76/359 with outline practical exercises and self-test questions. The level is carefully aimed at technician readers, assuming no great proficiency in mathematics, and with regard to the requirements of an up-to-date course.

CONTENTS: Physical quantities and electrical circuits. Capacitors and capacitance. Magnetic fields. Electromagnetism. Alternating voltages and currents. Reactive circuits. Semiconductor diodes. Transistors. Measurements.

96 pages June 1979 0 408 00433 9

MANUFACTURING TECHNOLOGY 2

P. J. HARRIS, Department of Engineering, Woolwich College

The work covered in this book is that required at Level II of the A5 mechanical and production engineering TEC programme. The unit for manufacturing technology (U76/056) has replaced an earlier unit (U75/035), but since some colleges still be working to unit U75/035 it was felt necessary to cover both units. The subject is treated from a practical viewpoint, but some mathematical treatment is considered, not just as an academic exercise but to illustrate how scientific concepts may be applied to efficient and economic manufacture.

CONTENTS: Heat treatment. Plastics. Presswork. Primary forming processes. Metal cutting. Machine tools. Measurement.

80 pages June 1979 0 408 00410 X